DEMCO

Animal Rights

Shasta Gaughen, *Book Editor*

Bruce Glassman, *Vice President*
Bonnie Szumski, *Publisher*
Helen Cothran, *Managing Editor*
David M. Haugen, *Series Editor*

Contemporary Issues
Companion

GREENHAVEN PRESS
An imprint of Thomson Gale, a part of The Thomson Corporation

Detroit • New York • San Francisco • San Diego • New Haven, Conn.
Waterville, Maine • London • Munich

LIBRARY OF CONGRESS CATALOGING-IN-PUBLICATION DATA

Animal rights / Shasta Gaughen, book editor.
   p. cm. — (Contemporary issues companion)
Includes bibliographical references and index.
ISBN 0-7377-2653-9 (lib. : alk. paper) — ISBN 0-7377-2654-7 (pbk. : alk. paper)
   1. Animal rights. 2. Animal welfare. 3. Livestock factories. 4. Animal experimentation—Moral and ethical aspects. I. Gaughen, Shasta. II. Series.
HV4708.A549 2005
179'.3—dc22
                                                            2004052352

Printed in the United States of America

CONTENTS

FOREWORD

In the news, on the streets, and in neighborhoods, individuals are confronted with a variety of social problems. Such problems may affect people directly: A young woman may struggle with depression, suspect a friend of having bulimia, or watch a loved one battle cancer. And even the issues that do not directly affect her private life—such as religious cults, domestic violence, or legalized gambling—still impact the larger society in which she lives. Discovering and analyzing the complexities of issues that encompass communal and societal realms as well as the world of personal experience is a valuable educational goal in the modern world.

Effectively addressing social problems requires familiarity with a constantly changing stream of data. Becoming well informed about today's controversies is an intricate process that often involves reading myriad primary and secondary sources, analyzing political debates, weighing various experts' opinions—even listening to first-hand accounts of those directly affected by the issue. For students and general observers, this can be a daunting task because of the sheer volume of information available in books, periodicals, on the evening news, and on the Internet. Researching the consequences of legalized gambling, for example, might entail sifting through congressional testimony on gambling's societal effects, examining private studies on Indian gaming, perusing numerous websites devoted to Internet betting, and reading essays written by lottery winners as well as interviews with recovering compulsive gamblers. Obtaining valuable information can be time-consuming—since it often requires researchers to pore over numerous documents and commentaries before discovering a source relevant to their particular investigation.

Greenhaven's Contemporary Issues Companion series seeks to assist this process of research by providing readers with useful and pertinent information about today's complex issues. Each volume in this anthology series focuses on a topic of current interest, presenting informative and thought-provoking selections written from a wide variety of viewpoints. The readings selected by the editors include such diverse sources as personal accounts and case studies, pertinent factual and statistical articles, and relevant commentaries and overviews. This diversity of sources and views, found in every Contemporary Issues Companion, offers readers a broad perspective in one convenient volume.

In addition, each title in the Contemporary Issues Companion series is designed especially for young adults. The selections included in every volume are chosen for their accessibility and are expertly edited in consideration of both the reading and comprehension levels of the audience. The structure of the anthologies also enhances accessibility. An introductory essay places each issue in context and provides helpful facts such as historical background or current statistics and legislation that pertain to the topic. The chapters that follow organize the material and focus on specific aspects of the book's topic. Every essay is introduced by a brief summary of its main points and biographical information about the author. These summaries aid in comprehension and can also serve to direct readers to material of immediate interest and need. Finally, a comprehensive index allows readers to efficiently scan and locate content.

The Contemporary Issues Companion series is an ideal launching point for research on a particular topic. Each anthology in the series is composed of readings taken from an extensive gamut of resources, including periodicals, newspapers, books, government documents, the publications of private and public organizations, and Internet websites. In these volumes, readers will find factual support suitable for use in reports, debates, speeches, and research papers. The anthologies also facilitate further research, featuring a book and periodical bibliography and a list of organizations to contact for additional information.

A perfect resource for both students and the general reader, Greenhaven's Contemporary Issues Companion series is sure to be a valued source of current, readable information on social problems that interest young adults. It is the editors' hope that readers will find the Contemporary Issues Companion series useful as a starting point to formulate their own opinions about and answers to the complex issues of the present day.

INTRODUCTION

In 1975, bioethicist and philosopher Peter Singer published the book *Animal Liberation: A New Ethics for Our Treatment of Animals.* In the preface, Singer proposed that people hold animals in a state of tyranny. "This tyranny," he wrote, "has caused and today is still causing an amount of pain and suffering that can only be compared with that which resulted from the centuries of tyranny by white humans over black humans. The struggle against this tyranny is a struggle as important as any of the moral and social issues that have been fought over in recent years." The publication of *Animal Liberation* was met with a great deal of controversy and had a profound effect on the emerging debate about issues such as how animals should be treated and whether or not animals were entitled to the same consideration and rights as human beings. Singer's arguments about the tyranny of humans over animals quickly became the basis for fledgling animal rights movements in the United States and Europe.

Singer's argument for animal liberation suggested that people needed to expand their moral horizons when it came to their thinking about animal treatment. The movement for animal liberation, Singer contended, was in many ways no different from the fight to end racial or sexual discrimination. "I believe that our present attitudes to these beings are based on a long history of prejudice and arbitrary discrimination," he wrote. Ultimately, attitudes toward animals "that were previously regarded as natural and inevitable come to be seen as the result of an unjustifiable prejudice." This was the primary goal of Singer's animal liberation movement: that animals be given the same consideration as any other species, including humans. Failing to give animals equal consideration, according to Singer, results in "speciesism," which he defines as "a prejudice or attitude of bias in favor of the interests of members of one's own species and against those of members of other species."

The modern animal rights movement that was prompted by Singer's influential work in *Animal Liberation* rests on four basic ethical principles. Singer outlined these principles as follows in the introduction to his book *Writings on an Ethical Life*, published twenty-five years after *Animal Liberation:*

> 1. Pain is bad, and similar amounts of pain are equally bad, no matter whose pain it might be. . . . This does not mean that pain is the only thing that is bad, or that inflicting pain is always wrong. Sometimes it may be necessary to inflict pain and suffering on oneself or others. . . . But this is justified because it will lead to less suffering in the long run; the pain is still in itself a bad thing.

2. Humans are not the only beings capable of feeling pain or suffering. . . . Of course, the nature of the beings will affect how much pain they suffer in any given situation.

3. When we consider how serious it is to take a life, we should look, not at the race, sex, or species to which that being belongs, but at the characteristics of the individual being killed, for example, its own desire about continuing to live, or the kind of life it is capable of leading.

4. We are responsible not only for what we do but also for what we could have prevented. . . . We should consider the consequences both of what we do and of what we decide not to do.

These four principles rest on the notion that animals, like humans, have sentience—that is, the capacity to experience both suffering and pleasure. Singer concludes that "if a being suffers there can be no moral justification for refusing to take that suffering into considera- tion. No matter what the nature of the being, the principle of equality requires that its suffering be counted equally with the life suffering . . . of any other being."

The influence of *Animal Liberation*, as well as subsequent writings by Peter Singer, has been significant for the modern animal rights movement. Singer himself is the president of Animal Rights Interna- tional and cofounder and president of the Great Ape Project, an orga- nization that seeks to obtain basic rights for chimpanzees, gorillas, and orangutans. Other organizations, such as the American Society for the Prevention of Cruelty to Animals (ASPCA) and the Humane Society of the United States (HSUS), fight to end unnecessary suffer- ing and exploitation of animals by humans. These organizations and others like them believe that the relationship between humans and animals should be guided by the four principles of ethical treatment outlined by Singer.

However, some animal rights organizations believe that Singer's principles do not go far enough. As Gary L. Francione, professor of law and Nicholas de B. Katzenbach Scholar of Law and Philosophy at Rutgers University Law School in Newark, New Jersey, writes, Peter Singer's philosophy of animal liberation does not provide an adequate basis for extending rights to animals. He explains, for example, that "although Singer opposes *most* animal experimentation, he does so because he thinks that most animal experiments do not produce ben- efits that are sufficient to justify the animal suffering that results. But he does not—and cannot—oppose *all* animal experimentation; if a particular use would, for example, really lead directly to a cure for a disease that affected many humans, Singer would approve that ani- mal use." Francione sees Singer as a proponent of animal welfare—the movement to treat animals humanely and compassionately—not as a proponent of animal rights, which seeks to abolish all human uses of animals.

The animal rights perspective (as opposed to the animal welfare position) is perhaps most visibly represented by the organization People for the Ethical Treatment of Animals (PETA). According to Harold H. Guither, emeritus professor of agricultural policy at the University of Illinois, PETA "shows the most rapid growth and influence of any animal rights organization" and boasts over three hundred thousand members. PETA founder Ingrid Newkirk has famously stated that "a rat is a pig is a dog is a boy," illustrating PETA's view that all animals should be considered equal under all circumstances. Newkirk believes that "animals have a worth in and of themselves, that they are not inferior to human beings but rather just different from us and they really don't exist for us nor do they belong to us." Expanding on this idea, PETA's philosophy boils down to a single creed: Humans do not have the right to use animals at any time, for any purpose. This philosophy conflicts with the philosophy of more moderate animal rights organizations. These groups are interested in the humane treatment of animals but are not fighting to completely eliminate the use of animals by humans.

The modern animal rights movement has grown and changed since it was sparked by the publication of *Animal Liberation*. The movement remains vibrant, and the fierce debate over how—and if—humans should use animals continues unabated. *Contemporary Issues Companion: Animal Rights* explores some of the most salient issues in the animal rights debate today, including whether or not animals should have the same rights as people, the debate over animal rights in the food industry, and the ethics of animal experimentation.

CHAPTER 1

SHOULD ANIMALS HAVE THE SAME STATUS AS PEOPLE?

Contemporary Issues
Companion

SHOULD ANIMALS HAVE RIGHTS?
A DEBATE

Roger Scruton and Andrew Tyler

Roger Scruton is a writer and philosopher. He is the author of
numerous books, including *Animal Rights and Wrongs*. Andrew
Tyler is the director of Animal Aid, an animal rights group in
England. In this selection, the authors debate the question of
whether animals have rights. Scruton takes the position that,
while humans should treat animals with care and respect, animals are not entitled to the same natural, moral rights as people.
He asserts that humans should strive to do as much good as possible in their role as the stewards of the earth, which includes
using animals for human good. Tyler, on the other hand, argues
that humans exploit animals at their convenience without true
regard for their welfare. Humans' poor stewardship of the earth,
he goes on, has resulted in environmental degradation and social
inequality around the globe. Tyler argues that humans should
acknowledge the moral rights of animals.

Dear Andrew Tyler

Many people now take the view that the human species is not
entitled to the dominion that it has so far asserted over all other
species. They express this by saying that animals, like us, have rights.
Hence many of the things that we do to animals are morally indefensible. I find myself agreeing with the conclusion, but not with the
premise. The attribution of rights to animals seems to me to be a radical departure from the norms of moral argument; if taken seriously it
would undermine our ability to make the important decisions that we
now must make if animals in general, and wild animals in particular,
are to enjoy a sustainable future.

The debate is not a trivial one. Advocates of animal rights are currently attempting to bankrupt a firm (Huntingdon Life Sciences),
which uses animals for medical research; they have succeeded in ban-
ning fur farming in Britain, and are now hopeful that they can ban
hunting with hounds. They intend, if successful, to ban shooting and

Roger Scruton and Andrew Tyler, "Do Animals Have Rights?" *The Ecologist*, vol. 31,
March 2001, p. 20. Copyright © 2001 by MIT Press Journal. Reproduced by permission.

angling, and no doubt there are those among them who would like to impose a strict regime of noninterference in the entire animal kingdom, whether the rest of us want it or not.

This intransigence is an inevitable result of the belief in rights. If I believe that you are denying someone his rights—to life, limb, property or freedom—then I am absolutely entitled to interfere on the victim's behalf. Rights may be relinquished—but only by the person who possesses them, and only if his action is entirely voluntary. The purpose of the concept of a right is to establish, around each individual, a sphere where that individual alone is sovereign. Hence your right is my duty, and if I disregard your rights I both wrong you and also do what is wrong.

Humans Have Sovereign Rights

Why should we have such a concept? Surely, because we wish to live in a condition of mutual freedom and mutual respect. The concept of a right derives from legal ways of thinking, and serves as the individual's shield against oppression. All calculation stops at the threshold where you are sovereign, and it is to mark out this threshold that we deploy the concept of a right. Some philosophers believe that there are both positive rights—which are laid down by a legal code—and natural rights—which are inherent in our condition as rational agents. And it is this idea of a natural right that is invoked by those who argue for the rights of animals. Natural rights are those like the rights to life, limb and freedom, the violation of which is tantamount to a declaration of war.

Let us suppose that animals do have rights; what follows? Surely, the very least that follows is that it is wrong to kill them, to eat them, to keep them as pets, to make them suffer in any way that is not to their individual benefit—and wrong in just the way that it is wrong to do any of this to a human being. That is what the activists say they believe. But do they really believe it? Are they prepared to say that my attempts to rid my barn of rats are tantamount to mass murder? That people who keep cats are complicitous in serial killing? That my keeping a horse in his stable is a case of false imprisonment? That my digging the garden involves the negligent slaughter of innocent worms, beetles and moles? Which activities involving animals would be permitted, and on what grounds?

But there is a more important consequence of rights-talk from the environmental point of view. To invoke rights is to accord absolute respect to the individual, and to give him precedence over collective calculations whenever his vital interests are at stake. Hence the sick, the deformed, and the genetically impaired have just the same rights as the healthy and the strong. If animals have rights, you have no more right to kill a sick, wounded or genetically impaired individual than you have to kill its healthy companion. All attempts at managing wildlife populations by encouraging healthy breeding and eliminating the carriers of diseases would be ruled out on moral grounds. It would also be morally impossible to intervene in nature to re-establish the ecological balance—say by culling an over-abundant predator population, by controlling parasites and pests, or by capturing animals and moving them to favourable breeding grounds.

Of course, if we lived in virgin forests as hunter-gatherers (itself morally impossible for the animal rights activist), we could reasonably assume that the ecological balance would restore itself over our footsteps. But we do not live like that. The environment is now our concern, something to be managed and restored by human ingenuity, and no longer able to restore itself unaided. To believe in the rights of animals we should have to relinquish that task, and allow animal populations to find what niche they can in the human sprawl. Good for rats and crows perhaps; but not for apes or fish or songbirds.

Roger Scruton

Support for Animal Rights

Dear Roger Scruton

You betray some panic over the inroads made by animal advocates. You complain that taking seriously the concept of animal rights represents a radical departure from 'the norms of moral argument', as though those often sterile, centuries-old 'norms' adequately serve us now . . . or ever did. Your notion of the absolute sovereignty of the human individual is a picture book romance, a conceit that has as much substance as candyfloss. More on this later.

You're panicked about the campaign successes relating to Huntingdon Life Sciences (a morally, scientifically and near-financially bankrupt company); about the fur farming ban; and about further ambitions in the direction of bloodsports. These successes do not arise because animal rights people have doped the nation's drinking water—but out of a rational objection to pointless animal abuse and a rejection of the flimsy justifications offered by those who orchestrate those abuses. The successes have come, notwithstanding a national media that, in the main, ardently promotes the status quo and which characterises effective dissent as mad and dangerous.

In such a climate, animal rights campaigners cannot possibly impose anything on the majority. Yet recent opinion polls show that most people oppose bloodsports, that vegetarianism is increasingly popular and that even opinion about animal experiments is finely balanced, despite ceaseless promotion of vivisection by the powerful scientific establishment, backed by government.

These poll showings indicate incipient, rather than grounded, support for the rights argument. It is the ambition of people like myself to encourage people into thinking further and acting accordingly—for example, into foregoing the enfeebled products of intensive animal farming and voicing opposition to vivisection—a practice that is a scientific nonsense, as well as an abuse of power, given that research data obtained from animals cannot be reliably applied to human beings.

Trampling Human and Animal Rights

I referred earlier to your invocation of the absolute sovereignty of the human individual. You say that this 'sovereignty' can be relinquished only voluntarily by the holder and that we are collectively bound (as individuals) to defend each other's rights. Where in the world, ever, has this seminar-room notion ever been played out? Western armed forces recently bombed Baghdad, killing many innocents, because Saddam's regime wouldn't accede to inspections of its weapons facilities. I publicly objected. Did you? The street children of Rio are shot dead, with government complicity, because they interfere with the tourist trade. A multinational pharmaceutical company arranged for a number of children to stand in a field in Egypt and be sprayed with an experimental pesticide so that it could collect toxicity data.

These are not unrepeatable travesties. This is the way of the world. The status quo is corrupt. The mighty trample the weak. Greed is rewarded. Decent sentiments are ever present but they are not sufficiently nurtured and celebrated by those at the commanding heights.

Yes, the notion of rights for human beings does exist thanks to centuries of struggle by oppressed groups. But while they are sacrosanct in the minds of some philosophers and in worthy pan-national declarations, they truly exist only for so long as, and to the extent that, the subject groups can defend them.

Though they are deserving of them, non-human animals have precious few rights, even in theory. In law they are property. There are certain rules about how we may exploit, kill and consume animals. But protection for animals stops at the point where we, as the exploiting species, are seriously inconvenienced or deprived of profit or pleasure.

Ending Animal Exploitation

The modern animal rights movement is concerned to address this situation and demonstrate that, just as commercial, cultural and intellectual life thrived in Britain after we gave up trading in Africans, so there is a positive future without exploiting animals. You must appreciate that ours is a practical movement. Our heads are not up our fundaments. We seek, first of all, to demonstrate what is often denied: that hundred of millions of animals every year in Britain alone are unwarranted victims of human commerce and culture—and that this occurs because animals have the status of mere objects. We promote ways of living that eschew the flesh, secretions and skins of animals. We repudiate the trade in 'pets' (give a home to a sanctuary animal instead!) and oppose leisure pursuits that depend on harrying and/or demeaning other species. While it is impossible in this imperfect world to live an immaculate life, the first principle should be: do as little harm as possible.

Let me now deal with this nonsense about 'human ingenuity' being needed to ensure a sustainable future.

This has come to mean scapegoating a range of 'alien' or commercially inexpedient species (from grey squirrels to badgers) for our own reckless excesses. The record shows that we fail when we try to poison and shoot our way to environmental harmony.

The record also shows that the only rational approach to take is to curb our own destructive appetites and to concede actual territory to non-human animals—not least to those from whom we are able to extract no obvious, or immediate advantage.

Andrew Tyler

Animals Do Not Have Morality

Dear Andrew

The notion of a right, I suggested, is an expression of the sovereignty that human beings claim over their own lives, and is only doubtfully applied to creatures who do not understand moral ideas, and who have no conception of their duties. You dismiss this as a 'seminar-room notion', without, however, proposing any alternative definition. The only intellectual enlightenment that I can glean from your response is that animals have rights because you say so. But you don't tell me which rights, which animals, or what this permits or forbids us to do to them. I would ask you again whether the killing of a rat is murder? If not, why not?

You seem keen to attribute 'panic' to me, which makes it sound as though you were offering not arguments but threats. I am far from disputing the claim that people abuse animals, but if this is to give grounds for condemnation we need to be much clearer than you are prepared to be about the nature of moral judgement. You ridicule my reference to the norms of moral argument, but your reply would be a mere tissue of self-serving emotion if it did not have a moral base. By attributing rights to animals you are making a moral judgement: but I contend that you are making it in the wrong terms, and that the consequences of doing so are intellectual and moral confusion.

Sustainable Management

You say that it is 'nonsense' to suggest that human ingenuity is needed to ensure a sustainable future. But, however badly we have managed things up to now, there is still no doubt in my mind that the answer is not no management at all but better management. I believe that the animals on our farm are in a more sustainable equilibrium since we shot the magpies and crows that were eating the young of other birds, that the cows are more healthy since we got rid of the rats (not by poisoning, hasten to add), that the voles and the carp have a better deal since the mink-hounds came. You may disapprove of these forms of management, but without them there would be very little bio-diversity here above the level of insect-life, and that too needs management. Even in your Vegan utopia, crops would have to be protected from pests, and bio-diversity maintained by intervention.

Of course, if animals have rights, then everything I have just described is morally wrong—indeed, morally impossible. You may think that, but since you have given no argument for it, and since I believe that the consequences of laissez faire would now be bad not only for us but for animals too, I reject your conclusion. It is good to feel sympathy for animals, to protect them and to do what you can to provide for their needs. But this does not alter the fact that it is we who do these things, because we live in a world that we control. To relinquish that control is merely to opt out of our principal responsibility. And it will not benefit the animals, even if it makes us feel good about them.

Of course animals are not 'mere objects'. They are sentient beings, towards whom we have real obligations. I sympathise with your aversion to the trade in pets, partly because I think it encourages people to see animals as persons, and so to degrade both themselves and their pets. And like you I would like to see more territory conceded to the animals. But this conceding of territory would be a form of management too, with people making all the important choices, and interfering to maintain things when pests, diseases or forest fires become a threat.

Roger Scruton

Freedom from Exploitation

Dear Roger

How odd! I set out an unequivocal case for non-violence against people and animals and you suggest I'm offering you 'threats'. Perhaps you've been reading too many *Daily Telegraph* leader columns— or writing them.

The panic I detected seemed to be rooted in your unease at the way the world is changing beneath your feet. The old conceits are fast unravelling—eg that killing animals for pleasure, money, or out of expediency is somehow a noble project.

The rights I would accord non-human animals are the rights not to be killed by humans—except unwittingly or in self-defence. Also, freedom from torture and exploitation. Those rights would extend to all species, which on the balance of the evidence, are sentient. I'm not concerned to judge and police other species in regard to their dealings with each other. Those who attempt such things have shown they have neither the requisite competence nor the vision. Besides, while some animals might kill others, they don't have our species' capacity for grandly choreographed, industrialised destruction.

Is killing a rat murder? Rats usually proliferate because people provide unguarded food sources. I know many animal rights people who farm, run factories and other businesses and are able to discourage unwelcome animals by non-violent means. Killing is lazy and nearly always futile: the vacuum left by the slain animal is filled by another, unless preventive measures are taken. So yes, killing a rat is unwarranted and immoral. But apply the term murder and, because of cultural conditioning, people react rather than think. So I wouldn't use it. But consider this: if someone encourages another to kill my child, most people would see that as aiding and abetting a murder. But what if a person publicly endorses a policy of aerial bombing that kills children (all of them sovereign, according to your formulation) thousands of miles away. Is that also aiding and abetting? My answer is yes, but distance and cultural bubblewrap get in the way.

"Do as Little Harm as Possible"

The case I've set out rests on a simple guiding principle, which I've already enunciated: do as little harm as possible. That is the test I would always apply, whilst appreciating that none of us can tread the earth without doing some damage.

The case you've set out against animal rights, it seems to me, rests on a pernicious conceit. It is that the world's flora and fauna benefits from human beings' 'stewardship'. The stewardship you favour involves identifying various problem species and slaughtering them; and identifying other species as having utility (cattle, sheep, cows, chickens, pheasants) and producing them in vast numbers, then killing them for food or sport.

The world resulting from such stewardship is polluted, over-heated, and unequal. It is locked into an unsustainable food production system that produces diseased animals and has seen the death of forests, the uprooting of thousands of miles of hedgerow, the prolific use of toxic chemicals and the imperilling of numerous species of songbird, mammal and fish, as well as a host of wildflowers and other plant life.

One of your favourite themes is that animals are not part of a 'moral community'. They are undeserving of rights because they have no duties or responsibilities. Clearly, you have much to learn about the richness of animal culture—of animals' powerful social bonds, their capacity for grief and sensual pleasure, their ability to think strategically and work cooperatively. The magpie feels no duty towards you, but then why should he? You want him dead.

Andrew Tyler

"Do as Much Good as Possible"

Dear Andrew

Rights protect the individual, not the species. By attributing rights to animals, therefore, we tie our hands when it comes to the great environmental questions that now confront us. You think that our hands ought to be tied: since animals have rights, then all culling, even of the diseased and the dangerous, is morally impossible.

You believe that animals have the 'right not to be killed by humans', but you do not judge and police other species. Presumably therefore you don't believe that a mouse has a 'right not to be killed by a cat'. I take this to show that all this talk of rights is so much hot air. What you are really describing is the duties of humans, not the rights of animals.

You refer to the 'richness of animal culture', and (in certain cases) this is an apt description—even of the rats in our ditches (which do not proliferate because we leave unguarded food, as you would like to believe). But that does not imply that animals organise their lives by moral principles, or that they blame and condemn each other as we do (and as you do preeminently), or that they settle their disputes by judgement and law. It is because we respond to the call of duty that we can be, and ought to be, stewards of our environment, and not just one competing species among many.

The world is polluted and over-heated not because of stewardship but because of bad stewardship. If everybody thought as you do, this would, I believe, simply entrench the position of humanity as the dominant species, while leaving all others to fend for themselves in a constantly dwindling habitat. It is fair enough to live by the principle 'Do as little harm as possible', in a world that is in a state of ecological balance. But that is not our world. I would prefer to say, when it comes to questions of the environment, 'Do as much good as possible'. That may mean also doing harm, and not only to rats.

Roger Scruton

Animal Rights Mean Social Progress

Dear Roger

You equate culling with 'stewardship' and still fail to acknowledge the mess we've made in this area. You suggest, instead, that all will be remedied by a bit of sensible tweaking.

Fewer mice—and birds—would be killed by cats if cats weren't mass produced for the pet industry. I don't believe, in any case, that you are remotely concerned about crimes against mice by cats. Your concern is that the extension of rights to animals would mean a curtailment of the freedom that people of your disposition currently enjoy to exploit and kill.

I clipped a book review you wrote for *The Times* in December 1996. The volume, by a 'pious Catholic' and public school Classics master, described the author's 'sporting' exploits in Cumbria. You wrote approvingly of his 'defiant celebration of the act of killing fish and birds in quantities that far surpass his gastronomic capacity . . . I find nothing strange' you declared, 'in the fact that these activities should be the high point of someone's life, and the object of powerful religious feelings.'

You tell me that your objective is to 'do as much good as possible'. Is this an example?

The movement for animal rights is another chapter in the history of social progress movements that has included the abolition of human slavery and the enfranchisement of women and the 'lower orders'. That Animal Aid recognises the continuity of such struggles is evidenced by the Living Without Cruelty Exhibition we staged in March 2000. Oxfam, Amnesty [International], World Development Movement, and Fairtrade Foundation were among the participants.

The animal rights movement is inclusive of anyone whose goal is a new, fairer deal for animals. Our view is that it is better to go part of the way towards a cruelty-free lifestyle than obstruct the way forward with blinkered and inconsistent hostility.

Andrew Tyler

ARGUMENTS FOR ANIMAL RIGHTS WILL LEAD TO FURTHER ABUSE OF ANIMALS

Damon Linker

Damon Linker is the editor of the journal *First Things*, which focuses on the role of religion and philosophy in the ordering of society. In this selection, Linker critiques the philosophical arguments for animal rights. Discussing the opinions of influential animal rights authors such as Peter Singer and Steven Wise, Linker explains the nuances of the different theories of the animal rights movement. Singer, for example, argues that because animals can experience pain and suffering, humans have no right to inflict pain on them. Wise goes further than Singer, Linker contends, by claiming that animals share autonomy—the ability to make decisions—with humans. Thus, according to Singer, not only should humans not inflict pain on animals, they must treat animals with the same consideration as fellow humans. Linker concludes that arguments that seek to equate animal concerns with human needs will result in more injustice toward animals, not less.

Not so long ago, animal-rights activists were viewed as crackpots if not thugs, the sort of people who splattered the fur coats of unsuspecting pedestrians with red paint or vandalized university research laboratories. This image was greatly enhanced by the extreme language in which the movement's leading figures routinely expressed themselves. Ingrid Newkirk, the longtime president of People for the Ethical Treatment of Animals (PETA), once opined that "a rat is a pig is a dog is a boy"; on another memorable occasion, she offered the thought that whereas "six million Jews died in concentration camps, . . . six billion broiler chickens will die this year in slaughterhouses."

In recent years, however, as some of the movement's most militant groups have moderated their rhetoric and their tactics, the cause of animal rights has begun to achieve a quite astonishing degree of respectability. On Capitol Hill, elected officials from both political par-

Damon Linker, "Rights for Rodents," *Commentary*, vol. 111, April 2001, p. 41. Copyright © 2001 by the American Jewish Committee. Reproduced by permission of the publisher and the author.

ties have become receptive to the movement's concerns, sponsoring legislation and even forming a congressional caucus called the "Friends of Animals." Perhaps even more strikingly, some of the nation's leading law schools, including Harvard, Georgetown, and Rutgers, have begun to offer courses in animal-rights law, and a Washington firm founded by two veterans of Ralph Nader's consumer-advocacy group has begun regularly filing suit to expand legal protections for animals. Its most notable achievement is a landmark federal-court ruling in 1999 that for the first time gives an individual the legal standing to sue on behalf of a distressed creature.

A Victory for Animal Rights

But the biggest victory so far for the animal-rights movement came [in] October [2000], when the U.S. Department of Agriculture (USDA) agreed to settle a lawsuit filed by a group called the Alternatives Research and Development Foundation that was seeking to expand the scope of the Animal Welfare Act of 1966. Since its passage, this act had been interpreted as empowering the USDA to oversee the treatment in laboratory experiments of large animals like dogs, cats, and primates. Left out of this regulatory regime were birds and, most importantly, mice and rats, which account for 95 percent of all animals used in scientific tests.

Under October [2000's] settlement—and despite the protests of indignant researchers—rodents are now to receive equal rights. The USDA has agreed to require universities, pharmaceutical companies, and other organizations that conduct biomedical experiments to fill out a report on the treatment of each and every warm-blooded animal, and to conduct random inspections to check for conformity to the new standards.

From the eagerness of politicians to please pet-owners to the all-too-familiar activism of the courts, one might point to all sorts of explanations for this dramatic change in the fortunes of the animal-rights movement. But the most basic reason is simpler and more ominous: many Americans have begun to accept the activists' argument that there is a moral imperative to treat animals like, well, people.

The Argument for Animal Rights

The first leg of this argument concerns the capacity of animals to feel, and the definitive treatment of the question may be found in a book that helped launch the modern animal-rights movement. This was *Animal Liberation* (1975) by Peter Singer, the Australian professor of philosophy who, amid much controversy, was recently appointed to a prestigious chair in ethics at Princeton.

The power of Singer's book derives largely from the simplicity of its argument. Following the lead of the 19th-century utilitarian philosopher Jeremy Bentham, Singer begins by identifying what he takes to be an indisputable moral intuition: we have an obligation not to inflict pain and suffering on creatures that are capable of experiencing them. Next, using a series of shockingly gory anecdotes, he shows that animals can and do experience these sensations. Finally, and completing his syllogism, he concludes that human beings must not inflict pain and suffering on animals.

Much of *Animal Liberation* is concerned with applying this principle to various human practices—from the use of animals in scientific experiments to raising and slaughtering them on "factory farms"—and then advocating the radical reformation or abolition of those practices. As Singer writes with characteristic bluntness, "there can be no reason . . . for refusing to extend the basic principle of equality of [ethical] consideration to members of other species." In this scheme, the dominion of humans over animals resembles nothing so much as a "tyranny" based on "arbitrary discrimination."

Indeed, in a parallel that shows up repeatedly in his subsequent writings and those of his many admirers, Singer compares the amount of pain and suffering inflicted on animals to "that which resulted from the centuries of tyranny by white humans over black humans." Just as slavery was based on racism, the abusive treatment of animals rests on what he calls "speciesism"—a prejudice "in favor of the interests of members of one's own species." Those who ignore the suffering of animals rely on the same sort of "difference that the most crude and overt kind of racist uses in attempting to justify racial discrimination."

For Singer, in short, only a complete break from the anthropocentric views of Western philosophy and religion will allow us to see that when it comes to the all-important capacity to experience pleasure and pain, we are morally indistinguishable from many of our fellow creatures.

Animals Are Like Us

Steven Wise's *Rattling the Cage* (2000) is perhaps the best-known book representing a relatively new tack in the project begun by Singer. He, too, offers an anguished cry of protest, again equating the present-day treatment of animals with human enslavement and describing the use of animals for medical research as "genocide." But for Wise, who teaches the new course in animal-rights law at Harvard and is a tireless courtroom advocate for the cause, the essential similarity between men and (some) beasts is autonomy: that is, the shared ability to form preferences and act on them.

Scientists, Wise points out, have demonstrated important physiological parallels between human beings and the primates that are his chief concern in *Rattling the Cage*. Chimpanzees, for instance, not only share 98.3 percent of our genetic make-up, but they have similar brain structures as well. On the evidence of evolutionary theory, he declares, "as recently as 5 or 6 million years ago, humans, chimpanzees, and bonobos were the same animal."

As a result, Wise argues, chimpanzees and bonobos possess myriad attributes that, through a combination of ignorance and prideful prejudice, we normally associate solely with human beings. They "feel" and "think," exhibit "emotions" and live in "cultures," "understand cause-and-effect relationships among objects, and even relationships among relationships." More importantly, observation of these animals in captivity and in the wild reveals that their behavior echoes ours in striking ways. They seem to possess elementary self-awareness: they imitate one another, engage in deception and trickery, and use their own primitive languages to communicate basic information and emotional states. A chimpanzee named Lucy even regularly prepared tea for researchers and masturbated to pictures of naked men in Playgirl magazine.

Contrary to Western prejudices that trace back to the Bible and Aristotle, Wise concludes, animals—or at least chimpanzees and bonobos—are not simply "things" that can be treated as property. They are, instead, "persons" in the legal sense. Like us, that is, they are bearers of individual rights and deserve to be treated accordingly. As the renowned primatologist Jane Goodall puts it in her glowing foreword to *Rattling the Cage*, this book is meant to be the animals' "Magna Carta, Declaration of Independence, and Universal Declaration of Rights all in one."

Humans Are Unique

That such arguments have found an audience at this particular cultural moment is not so hard to explain. Our popular and elite media are saturated with scientific and quasi-scientific reports claiming to prove the basic thesis of the animal-rights movement. Having once believed ourselves to be made in the image of God, we now learn—from the human genome project, the speculations of evolutionary psychologists, and numerous other sources—that humankind, too, is determined by genetic predispositions and the drive to reproduce. We are cleverer than other animals, to be sure, but the difference is one of degree, not of kind. As Verlyn Klinkenborg wrote on the editorial page of the *New York Times*, "Again and again, after starting from an ancient premise of radical differences between humans and other creatures, scientists have discovered profound similarities."

But have they? Genetics and evolutionary biology may be, indeed, extremely effective at identifying the traits we share with other species. But chemistry, for its part, can tell us about the ways in which we resemble chunks of charcoal, and physics can point to fundamental similarities between a man and all the matter in the universe. The problem with these observations is not that they are untrue. It is that they shed no light whatsoever on, or rather they are designed to obfuscate, what makes humanity unique as a species—the point on which an answer to the likes of Peter Singer and Steven Wise must hinge. For his part, Singer commits the same error that John Stuart Mill found in the system of Jeremy Bentham: he makes no distinction among kinds of pleasure and pain. That animals feel emotions can hardly be doubted; but human beings experience life, even at its most "animalistic" level, in a way that fundamentally differs from other creatures.

Thus, Singer can account for the pain that humans and animals alike experience when they are hungry and the pleasure they feel when they eat, but he cannot explain, for example, a person's choice to starve himself for a cause. He understands that human beings, like animals, derive pleasure from sex and sometimes endure pangs of longing when they are deprived of it, but he cannot explain how or why, unlike animals, some choose to embrace celibacy for the sake of its noble purity. He is certainly attuned to the tendency we share with animals to fear and avoid pain and bodily harm, but he is incapable of understanding a man's willingness to face certain death on the battlefield when called upon to do so by his country. Still less can he explain why stories of such sacrifice sometimes move us to tears.

In much the same way, the evidence adduced by Steven Wise to suggest that primates are capable of forming rudimentary plans and expectations fails to demonstrate they are equal to human beings in any significant sense. Men and women use their "autonomy" in a world defined not by the simple imperatives of survival but by ideas of virtue and vice, beauty and ugliness, right and wrong. Modern scientific methods, including those of evolutionary psychology, have so far proved incapable of detecting and measuring this world, but that does not make any less real the experience that takes place within it.

Western civilization has tended to regard animals as resembling things more than human beings precisely because, like inanimate objects, and unlike the authors of the real Magna Carta, animals have no perception of morality. Until the day when a single animal stands up and, led by a love of justice and a sense of self-worth, insists that the world recognize and respect its dignity, all the philosophical gyrations of the activists will remain so much sophistry.

Putting Human Interests First

None of this, of course, exempts human beings from behaving decently toward animals, but it does provide a foundation, when necessary, for giving pride of place to the interests of human beings. This has particular relevance for biomedical research.

Among the most vociferous critics of the USDA's capitulation to the animal-rights movement were the nation's leading centers of medical science. The National Association for Biomedical Research estimated that the new regulations would cost universities alone as much as $280 million a year. Nor is the issue simply one of dollars. As Estelle Fishbein, counsel for Johns Hopkins University, recently argued in the

Journal of the American Medical Association,

> Genetic research promises to bring new therapies to alleviate human suffering from the acquired immunodeficiency syndrome, Parkinson's disease and other neurological diseases, and virtually all other human and animal diseases. However, the promise of this new era of medical research is highly dependent on the ready availability of mice, rats, and birds.

Far from being a mere administrative hassle, she concluded, the new regulations would "divert scarce grant funds from actual research use, distract researchers from their scientific work, and overload them with documentation requirements."

Serious as this threat is, a still more troubling one is the effect that the arguments of animal-rights proponents may have, in the long term, on our regard for human life itself. Peter Singer's appointment at Princeton caused a stir not because of his writings about animals but because of his endorsement of euthanasia, unrestricted abortion, and, in some instances, infanticide. But all of his views, as he himself maintains, are of a piece. The idea that "human infants and retarded adults" are superior to animals can only be based, he writes, on "a bare-faced—and morally indefensible—prejudice for members of our own species."

In much the same way, Steven Wise urges us to reject absolute demarcations between species and instead focus on the capacities of individual humans and individual apes. If we do that, we will find that many adult chimpanzees and bonobos are far more "human" than newborn and mentally disabled human beings, and thus just as worthy of being recognized as "persons."

Though Wise's inference is the opposite of Singer's—he does not wish to deprive underdeveloped humans of rights so much as to extend those rights to primates—he is playing the same game of bait-and-switch: in this case projecting the noblest human attributes onto animals while quietly limiting his sample of human beings to newborns and the mentally disabled. When raising animals to our level proves to be impossible, as it inevitably must, equal consideration can only be won by attempting to lower us to theirs.

Antihuman Prejudice

It is a curious fact that in virtually all of human history, only in liberal democracies—societies founded on the recognition of the innate dignity of all members of the human race—have animals enjoyed certain minimum protections, codified in our own country in the Animal Welfare Act. It is a no less curious fact that these same liberal democracies have become infected over the past decades with a corrosive self-doubt, giving rise in some educated circles to antiliberal, antiwhite, antimale, anti-Western, and now, with perfect logic, antihuman enthusiasms.

The proponents of these various but linked ideologies march under a banner of justice and the promise of extending the blessings of equality to one or more excluded others. Such piety is to be expected in a radical movement seeking well-meaning allies; but it need not deflect us from the main focus of their aggressive passions, which the euthanasia-endorsing Peter Singer, for one, has at least had the candor to admit to. Can anyone really doubt that, were the misanthropic agenda of the animal-rights movement actually to succeed, the result would be an increase in man's inhumanity, to man and animal alike? In the end, fostering our age-old "prejudice" in favor of human dignity may be the best thing we can do for animals, not to mention for ourselves.

ASCRIBING MORAL VALUE TO ANIMALS IS LEADING TO MORE ANIMAL RIGHTS

Eric Sundquist

The idea that animals may have the same moral value as human beings is becoming more popular, writes Eric Sundquist in the following selection. In the past, he explains, animals have been considered to be automata—lacking in the capacity for the thought and emotion that humans possess. However, researchers, philosophers, and animal rights activists have proposed that animals do indeed feel and suffer, Sundquist relates, and as a result, interest in animal welfare has exploded. The change in attitudes toward animals underlies movements that range from advocating tougher animal welfare laws to granting some animals the same legal rights as humans, Sundquist reports, and has also led several states—and countries—to adopt new animal rights laws. Sundquist writes for the *Atlanta Journal-Constitution*.

Florida voters [decided] in November [2002] to amend the state's constitution to protect pigs.

It's an unorthodox proposal, but proponents are deadly serious. Nearly a half-million Floridians signed petitions to get the measure on the ballot.

The amendment combat[s] so-called "factory farms," which confine thousands of pigs to cages so small that the animals cannot turn around. The measure prohibit[s] farmers from caging pregnant sows in this way, discouraging industrial-scale hog farms from setting up shop in the state. Voters consider[ed] animal protections in other states as well. While Floridians vote[d] on hogs, Oklahomans decide[d] whether to ban cockfighting, and Arkansas voters consider[ed] making some forms of animal cruelty a felony.

Across the Atlantic, Germany amended its national constitution [in 2002] to protect "the natural foundations of life" for animals as well as people. Switzerland adopted a constitutional amendment in 1992 acknowledging animals as "beings" rather than things.

Changing Values

In short, the Western world, which not long ago ascribed the same moral value to animals as to plants or stones, is having a change of heart:

• Thirty-seven states have made certain forms of cruelty to animals a felony; four such laws were enacted [in 2002] alone. Georgia's took effect in 2000 and resulted in its first conviction [in 2002], when an Atlanta teenager pleaded guilty to dousing a dog with gasoline and setting it on fire.

• Western religion has traditionally denied that animals have souls, but a plurality of Americans now apparently disagree. Forty-three percent say pets go to heaven, compared with 40 percent who say they don't and 17 percent who don't know, according to a 2001 ABC-News/Beliefnet poll.

• In the '80s and '90s, protectionists' protests made fur-wearing controversial and pressured many companies into lessening or halting product testing on animals. With stricter federal controls, the number of animals used in research experiments also declined. Every Western country but the U.S. has stopped experimenting on chimpanzees, for example, and the practice is increasingly rare here.

• Philosophers, who pretty much ignored the subject until the 1970s, have churned out innumerable books and articles weighing ethical issues relating to animals. These in turn have inspired political and personal action for "animal liberation," as Princeton professor Peter Singer's seminal 1975 book was titled.

• Law schools have taken up the issue as well, with more than two dozen in the United States now offering animal-law courses.

• Vegetarianism is in vogue. A 2002 *Time* magazine poll found that 10 million Americans are vegetarians. Following the lead of many sit-down restaurants, [in 2002] Burger King rolled out its new veggie burger across the nation.

Animal protectionists can seem like extremists barely clinging to the fringe: People for the Ethical Treatment of Animals [PETA] cavort in silly costumes or in the nude, while activists from the Animal Liberation Front commit vandalism or worse.

Yet these antics belie the depth and seriousness of the change of heart on animals. PETA, for example, is dwarfed in size by the less flashy but arguably more effective Humane Society of the United States [HSUS], whose mailing list of 7 million is larger than that of the National Rifle Association.

HSUS is behind many of the 21 pro-protection votes on citizen-sponsored, statewide ballot measures since 1990. (In contrast, there were only two such votes from 1930 to 1989.) The ballot measures, and the popular move among legislatures to increase criminal penalties for cruelty, reflect the growing belief that animals have intrinsic worth, and there are moral considerations in causing them to suffer.

Americans even seem to embrace the concept of legal rights for some animals. According to a 1999 Zogby poll for a chimp-advocacy group, 51 percent of Americans say chimps should have rights "similar to children with a guardian to look out for their interests," and 9 percent say they should have the same rights as adults.

Animals as Automata

Why the sudden interest in the welfare—and possible rights—of animals?

Through much of Western history, despite occasional dissents from the likes of St. Francis of Assisi and English political philosopher Jeremy Bentham, the consensus was that humans could pretty much do what they pleased with animals. This was Aristotle's belief—"animals exist for the sake of man"—and that of most Christians and Jews, based in part on God's covenant with Noah—"into your hand they are delivered."

In the 17th century, Enlightenment philosopher Rene Descartes argued that not only do animals lack souls, but they also lack thought and the ability to suffer.

In the 19th century, however, Darwin's theory of evolution undermined Decartes' view of animals as automata—though that realization has been slow to sink in. How could people descend from automatons? At what point did they suddenly begin to think, suffer and acquire souls?

More challenges came with findings that many animals have genetic makeups and nervous systems similar to ours. Now animal behavior research reveals that some animals can count, pass learned behavior to their young and even "speak" through sign language. The old view, that humans and animals are different kinds of beings, is harder to support.

New Considerations Regarding Animals

Other factors have prompted the reconsideration of animals as well:

• The civil rights movement broke the mindset that there were anything but arbitrary reasons to discriminate based on race. The women's movement did the same for gender discrimination, and later the gay-rights movement challenged the idea that sexual identity was grounds for mistreatment. Extending consideration to animals was, to some, a logical next step. Writes *The Color Purple* novelist Alice Walker, "The animals of the world exist for their own reasons. They were not made for humans any more than black people were made for whites or women for men."

• Environmentalists popularized the notion that it is wrong for humans to thoughtlessly exploit the world. Rachel Carson's *Silent Spring* of 1962 contained parallel themes of concern for the environment for animals: "As man proceeds toward his announced goal of the conquest of nature, he has written a depressing record of destruction, directed not only against the earth he inhabits but also against the life that shares it with him."

• Animal protectionists effectively publicized some ugly incidents involving use of animals. After receiving photos of animals suffering in dog-dealer and laboratory facilities, *Life* magazine published the 1966 article "Concentration Camps for Dogs." Later that year, Congress passed the Laboratory Animal Welfare Act. In 1984, PETA distributed a videotape showing researchers at the University of Pennsylvania inflicting head injuries on monkeys. The following year Congress strengthened the Animal Welfare Act, requiring that research institutions establish committees, with nonaffiliated members, to review protocols involving some species of animals.

Some Suffering Acceptable?

In response, ethicists, philosophers and theologians have begun to reappraise their views. From a conservative Christian perspective, for example, evangelist J.R. Hyland argues that Jesus' "cleansing of the Temple" was not an attack on the financial practices of the money-changers, as it's usually seen, but on the animal sacrifices the money-changers were facilitating.

Probably most influential have been the arguments of secular philosophers, notably Peter Singer, whose *Animal Liberation* has sold more than 500,000 copies and continues to be reprinted.

Humans' mental superiority is no defense for what some are now calling "speciesism," he argues. If it were, then it would be all right to discount the pain of people without full mental abilities—infants or the mentally retarded, say—and to perform painful experiments on them.

"What we must do is bring nonhuman animals within our sphere of moral concern and cease to treat their lives as expendable for whatever trivial purposes we may have," Singer wrote.

This does not mean that people may not use animals at all. Though Singer argues, for example, that the suffering of animals in modern factory farms does not outweigh the benefit to humans of eating meat when there are healthful vegetarian alternatives, he allows that there will be times when a balancing of interests allows animal suffering. "If one, or even a dozen animals had to suffer experiments in order to save thousands, I would think it right and in accordance with equal consideration of interests that they should do so."

Other animal protectionists go further. Tom Regan, a philosopher at North Carolina State University, denies that Singer's "welfarist" framework amounts to liberation at all. Singer's view, he says, "permits utilizing other animals for human purposes, even if this means (and it always does) that most of these animals will experience pain, frustration and other harms, and even if it means, as it almost always does, that these animals will have their lives terminated prematurely."

Regan holds that animals, like people, must not be used as means to ends—they have rights, not just interests to be balanced. Regan would abolish the use of animals for food, experiments or nearly any other purpose, a position echoed by PETA in its slogan, "Animals are not ours to eat, wear, experiment on, or use for entertainment."

Legal Personhood for Chimps

Protectionists, inspired by the ethical arguments of both Singer and Regan, have advanced various arguments for establishing legal rights for animals. One lawyer who commands attention inside and outside the movement, through two popular books and an association with primate researcher Jane Goodall, is Stephen Wise, who teaches at Harvard and Vermont law schools.

Wise points out that children and the mentally ill are viewed as persons with rights, but not all the rights of a competent adult. He argues that some animals also deserve legal personhood with proportional rights. Animals wouldn't be able to vote, but they might have rights to liberty or bodily integrity.

Humans, bonobos, gorillas, orangutans and Atlantic bottle-nosed dolphins clearly qualify for the granting of legal personhood, Wise argues in his 2002 book *Drawing the Line: Science and the Case for Animal Rights*. African gray parrots and African elephants may also qualify under a more expansive interpretation.

"It takes a while for these ideas to get a toehold in the legal system," Wise said in an interview. "If you've never seen these arguments before, they can seem strange."

He says that chimpanzees may get rights first, because of their high intelligence and low commercial value. They might be assigned legal guardians, much as parentless children are.

"Once that happens," he says, "there will be a paradigm shift. I predict there will be a gradual extension of rights."

Factory Farms

Until animals are accorded formal rights, protectionists continue to tackle one welfare issue at a time.

Though the last two decades have seen tougher anti-cruelty laws and restrictions on animal research, American agriculture remains largely exempt. In general, pigs and cows have less protection than lab rats and mice, even though pigs are considered at least as intelligent as dogs. Federal law governs some aspects of transportation and slaughter of livestock, but the industry has fought off almost all efforts to set standards for the actual raising of animals.

"There is essentially no regulation of the way animals are housed and kept," complains Wayne Pacelle of the Humane Society of the United States.

As a result, millions of animals spend their lives on "factory farms," confined in cages so small that they can barely move.

The European Union has taken steps to curtail or ban such practices as raising veal calves in small crates, keeping chickens nearly immobile in "battery cages," and holding pregnant sows in "gestation crates." Florida's ballot initiative addresses the latter; if it passes, it would be the only statewide ban on such cages.

It is opposed by agriculture. Doing away with intensive agricultural practices, argues Joe Miller of the American Farm Bureau, would result in one of two outcomes: "starvation, or higher prices." Miller acknowledges that hogs in gestation crates and other tightly caged animals cannot move enough to turn around, but that has some humane benefits as well as economic ones, he says: The animals are, for example, protected from predators and disease.

The numbers of animals involved in American agriculture are huge—some 8 billion are killed each year, according to the Humane Society. In comparison, hunters kill an estimated 100 million, researchers perhaps 20 million.

At such a mass scale, it may be hard to focus on individual animals. But that's the task protectionists have set for themselves. Says the Humane Society: "All of the animals we refer to as 'farm animals' have unique personalities. They're fascinating creatures with the ability to love, form friendships, mourn, get angry and show a variety of other emotions. They're deserving of our respect, our compassion and our gratitude for what they give us."

And maybe more legal protection, too.

NONHUMAN APES DO NOT HAVE HUMAN-LEVEL SELF-AWARENESS

Clive D.L. Wynne

In the following selection, Clive D.L. Wynne discusses whether apes—such as chimpanzees—are self-aware. This is an important question, Wynne points out, because chimpanzees are used in some biomedical research experiments. If they are self-aware—that is, if they suffer in the same way human beings do—many would argue that they should not be used in research. Wynne examines various ways of determining self-awareness in apes, such as language use, self recognition, and cognitive capacity, and concludes that apes do not exhibit the same level of self-awareness as humans. Wynne leaves open, however, what the ethical implications of this may be. Wynne is an associate professor in the psychology department at the University of Florida. He is the author of *Animal Cognition: The Mental Lives of Animals* and *Do Animals Think?*

My [article] title is borrowed from [the] book [*The Soul of the Ape*] by Eugene Marais (1871–1936), published posthumously in 1969. Marais was one of the first to make a close study of the behavior of nonhuman primates in the wild. His analyses were far ahead of their time, because he attempted to understand the actions of these creatures in terms of the evolutionary continuity with *Homo sapiens* that Charles Darwin had earlier proposed. Although genetic analyses were still a long way off, it was obvious to Darwin and his converts in the 19th century that the great apes—chimpanzees, gorillas, orangutans and bonobos, or pygmy chimps—must be our distant cousins. "He who understands baboon," Darwin noted, "would do more toward metaphysics than Locke."

Although today we recognize baboons as monkeys, not apes, and therefore less closely related to us, Darwin's point still stands. And Marais was on the right track when he observed free-living and captive baboons in his native South Africa and puzzled over the similarities—and differences—he could see between their behavior and that of human children and adults.

Clive D.L. Wynne, "The Soul of the Ape," *American Scientist*, vol. 89, March 2001, p. 120. Copyright © 2001 by Sigma Xi, The Scientific Research Society, Inc. Reproduced by permission of the publisher.

Rights for Nonhuman Apes

Most of the questions with which Marais grappled remain unanswered. But in recent years attempts to understand primate minds have spawned a controversial new ethic. It is based on the notion that the differences in psychology between people and other great apes are too small to justify different treatment. That is, nonhuman apes should get the same ethical consideration that their human brethren receive. This view is most manifest in the Great Ape Project, which describes itself as working to raise the legal and moral status of these animals. The ultimate aim of this group is to have the United Nations adopt a declaration on the rights of nonhuman apes, one that would make all medical research on them impossible.

Supporters of the Great Ape Project are impressed by the genetic similarity between people and nonhuman apes and by the relatively short period of time since we and our closest relatives (chimpanzees) diverged from a common ancestor. Most advocates of this project point to what they consider the key psychological similarity between nonhuman apes and ourselves: Nonhuman apes, they argue, are self-aware. As a consequence of this self-awareness, gorillas, chimpanzees, bonobos and orangutans must suffer in captivity in ways not so different from what we would experience under similar circumstances.

High Stakes

This is not an abstract scholarly debate. There are about 1,600 chimpanzees held for biomedical research in the U.S., and these animals are essential to the study of several maladies, ones for which few if any other approaches are available. Probably the single most important example is liver disease. It was research on chimpanzees that provided the vaccine against hepatitis B. Carriers of hepatitis B are about 200 times more likely to develop liver cancer than the general population, so the hepatitis B vaccine can be considered the first cancer vaccine. More than a million people in the United States have been infected with hepatitis B, and nearly half the global population is at high risk of contracting this virus. Chimpanzees have also been crucial for the study of hepatitis C, a chronic disease for some four million Americans.

And hepatitis just heads the list. AIDS is another prominent example, because chimpanzees are the only nonhuman species that can be infected with HIV-1, the common form of the virus found throughout the world. The reason became clearer [in 2000], when investigators found proof that HIV-1 spread to humans from chimps early in the 20th century. Now more than 36 million people around the world are infected with this virus, and some 22 million have already died from AIDS. Chimpanzees continue to be immensely valuable in the search for a vaccine. Chimps are also helping scientists to battle other dire health problems, including spongiform encephalopathy ("mad-cow disease"), malaria, cystic fibrosis and emphysema.

So the stakes are very high indeed. On the one hand, precisely because they are closely related to us, nonhuman great apes are the only suitable species for research on several diseases that cause immense human suffering. On the other hand, these creatures may be so similar to ourselves that making them the subjects of biomedical experimentation may be difficult to justify ethically. This could be a Faustian dilemma—are we selling chimp souls to better our human lives?

I don't pretend to have the answer to such a weighty moral question. But as a psychologist I do feel qualified to examine one of the reasons being cited for providing these creatures with rights: the idea that nonhuman great apes are self-aware. In my view, the evidence for this assertion is considerably exaggerated. Several lines of reasoning have been put forward to buttress this claim. Let me examine each of these arguments in turn.

Apes and Language

First, it is said, the minds of nonhuman great apes are much like ours because some of these animals have been taught to use language—sign language. Research on the abilities of chimpanzees and bonobos to communicate using signs has been going on for more than 30 years. The early excitement that accompanied reports of signing chimpanzees has died down, perhaps because the details are not that compelling. In all these studies, the animal's vocabulary developed painfully slowly, and it never exceeded a couple of hundred signs (about two weeks' work for a healthy two-year-old child). Often the chimp involved could only repeat gestures, and in any case, chimpanzee "sentences" rarely extend beyond one or two signs—making any discussion of grammar or syntax seem rather forced.

The occasional anecdote of a chimpanzee forming a novel combination of signs to represent a word he had not been taught (such as signing water-bird for swan) prompted some to believe that these animals indeed have an innate capacity for language. But in reality such cases are extremely rare and quite hard to interpret. How does one know, for example, that this chimp did not intend two distinct utterances, water and bird, to indicate the two things then in view? In my estimation, such curiosities contribute little to the debate.

Somewhat more relevant are recent reports of a bonobo named Kanzi, whose linguistic abilities (as expressed by pressing buttons on a special keyboard) are alleged to far exceed those seen during the earlier sign-language studies with chimpanzees. A critical test for Kanzi's comprehension of sentence structure involved asking him to respond to an instruction such as "would you please carry the straw?" Sure enough, Kanzi picks up the straw. But there's a weakness here. Although it could be that grammar conveyed the correct meaning of the test sentence, it could equally be just the circumstances that made the requested action obvious (given that Kanzi knows what the words carry and straw refer to). After all, a chimp may carry a straw; a straw cannot carry a chimp. Investigators have also claimed to discern grammatical structure in Kanzi's own keyboard strokes. But because the average length of his utterances is around 1.5 pushes, I believe that a meaningful analysis of grammar or syntax is impossible.

Self-Recognition in Apes

The second tier of evidence for self-awareness in nonhuman apes comes from so-called mirror tests. These experiments can be conducted in several different ways. For example, while the subject is anesthetized or sleeping, the experimenter places a mark on the forehead or ear with an odorless, tasteless dye. Upon waking, the animal is shown a mirror. Will the creature recognize that the splash of dye is on his own face? The answer is taken to be yes if the animal touches the marked area of skin more often with a mirror in front of him than without.

Although there were methodological problems with some earlier studies, it is now broadly accepted that chimpanzees, bonobos and orangutans can recognize themselves in mirrors. Claims that dolphins and gorillas can pass this test are disputed. And all other species examined (including fish, dogs, cats, elephants and parrots) react to themselves in a mirror either not at all or as if the reflection is another animal.

The problem with the mirror test of self-recognition lies not in the results—clearly some nonhuman apes can recognize themselves in mirrors—but in the interpretation. Why should such self-recognition be equated with self-awareness? Some people cannot recognize themselves in mirrors (blind people are the most obvious group), but nobody suggests they lack any aspect of self-awareness. Conversely, in autistics self-awareness is clearly impaired. And yet autistic children develop the ability to recognize themselves in mirrors at the same rate as normal children. So these tests of self-recognition in mirrors are interesting and doubtless say something about how animals view their bodies, but they tell us nothing about the deeper question of self-awareness.

Testing Cognitive Capacity

The final set of evidence for self-awareness in nonhuman apes comes from something called "cognitive perspective taking." Daniel Povinelli and his colleagues at the University of Southwestern Louisiana developed the best-known experiment of this sort. They based it on certain tests that are used to judge mental development in autistic children. In Povinelli's "guesser-knower" experiment, a chimpanzee watches a trainer put food into one of four cups. The chimpanzee knows that the cups are there (he was shown them before the experiment started) but cannot see which of them receives the food. Another trainer who has viewed the manipulations (the knower) then points to the baited cup. A third trainer who has not seen the food go in (the guesser) points to a different cup. A chimp—or child—who has an awareness that others have minds, can readily appreciate that one trainer knows where the food is hidden, and the other one doesn't.

Chimpanzees in these experiments were ultimately successful in picking the cup containing the food (selecting the cup pointed to by the knower), but they required many hundreds of training experiences before they could make the right choice reasonably consistently. This pattern suggests only that the apes involved gradually learned to associate one stimulus (the knower) with the reward, not that the animal is treating the trainers as people with minds.

In more recent experiments, Povinelli and his coworkers offered a chimpanzee a choice between begging from a person who could see various pieces of food and begging from a person who could not (because she was blindfolded, for example). To the experimenters' surprise, chimps were initially just as likely to approach the trainer who could not possibly see the food. With enough experience, the chimps gradually learned to ask only the person with unrestricted vision for a tasty morsel. But they showed no spontaneous understanding that being unable to see disqualified a person from providing snacks. So here again, I find little to indicate that apes have any awareness of the minds of others, much less that they have an awareness of their own thoughts.

In truth, scientists do not yet know much about the soul of the ape. But we've learned a few things since Marais made his pioneering observations. It is clear that these animals do not show self-awareness in anything like the ways a human does, even when given the tools to do so. But if their minds are not like ours, what are they like? I don't know. I don't even know how we would draw ethical conclusions from that knowledge if we had it. The only thing I'm sure of is that we should view apes as worthy of wonder for being what they are— not merely as reflections of ourselves.

Nonhuman Apes' Human-Level Self-Awareness Entitles Them to Legal Status

Julie Cohen

In the following selection, Julie Cohen, a writer for *Geographical* magazine, describes her surprise at discovering the advanced communicative abilities of primates. Surely, she writes, animals with this level of cognitive awareness deserve the same legal rights as human beings. Not all species should have the same rights, Cohen asserts, but legislation such as that enacted by New Zealand's Parliament on behalf of great apes seems to illustrate that these animals deserve legal protections. However, Cohen points out, as long as animals are regarded legally as property, making the case for legal rights will be difficult. Nevertheless, the author writes, test cases involving animals should lead to more rights for animals in the future.

Sitting on the forest floor opposite a haughty female ape I was completely taken aback when she tipped her head on one side and as she pressed the buttons on an electronic keyboard, a synthesised voice said, "Has the visitor brought a surprise?" Thankfully Bill, a researcher at the Georgia State University Language Research Centre came to my aid saying, "Yes, yes she has—she brought you some jello." "Good," came the satisfied response.

Before visiting Sue Savage-Rumbaugh and her talented team of researchers at the Georgia University Language Research Centre I would have been cautious to suggest animals deserved legal rights. I would have agreed that anti-cruelty rules are too weak and that animals need to be treated with kindness, respect and care, but legal rights for animals may have seemed too extreme. But spending an afternoon in the company of Panbanisha, a 14-year-old bonobo, or pygmy chimpanzee, and her one-year-old baby, Nyota, convinced me otherwise.

Legal Rights for Animals

For years animal rights groups have put forward the case for legal rights but it is only now that demands are being taken seriously. [In

2000] Harvard and Georgetown law schools began teaching a new course in animal rights law. That the two most prestigious law schools in the USA are taking this new field of law seriously gives an indication of the change in public feeling. Animals have always been regarded as property under both American and British law. But as Steven Wise, the charismatic lawyer teaching the new Harvard Law School course on animal rights law explains, we should challenge our conceptions.

"What is it that makes humans qualify for legal rights and not animals?" he asks. "In Greek and Roman times humans were deemed superior because they believed God had designed a hierarchical world where humans were dominant. Animals were for human use. They were regarded as mere property. That has stuck in the eyes of the law and yet it is a principle without intellectual foundation. Darwin showed that the world was created in a far more random way and that humans are not superior to anything by divine decree."

Wise points out previous misconceptions that have now been remedied. Women were considered inferior to men until relatively recently and for many years black people were not allowed equal rights to white people. This illustrates how wrong our conceptions have been in the past. If the way women and black people were treated had not been challenged, then the inequalities would have continued.

Communicating with Animals

Panbanisha is lucky. It is clear from her attitude she is in no doubt that she is a star and uses this celebrity status to her benefit as much as possible.

When the photographer asked to take her picture a delicate negotiation process had to be undertaken. "You're beautiful," he told her in very embarrassed whispers. She was pleased. "I'll bring you a drink," he offered seeing that he was getting her interest. "Coffee, milk and juice with ice," she demanded, clearly delighted.

From that moment on she had the photographer and I obeying her every whim. Before she would answer questions about what she likes to eat and do she insisted on playing hide-and-seek in the woods belonging to the centre.

The simple questions I asked her were clearly way below her capacity as she demonstrated when she astounded me by expressing her sadness that Kanzi, a male bonobo she lives alongside, was upset at being left out.

Who would have considered that an animal was capable of expressing feeling for another animal? Panbanisha demonstrated that we should not judge an animal's consciousness by their language ability. That they may not be able to talk does not necessarily mean they can not understand or think.

At the Arizona State University, Irene Pepperberg, an associate professor in the department of ecology and evolutionary biology, has been working with an African Gray Parrot named Alex, who has learnt to identify more than 71 objects, actions, colours, shapes and materials. If he sees someone pick up a cup of tea he will warn them with "Hot".

This, critics could argue, could all be just rote learning. A few years ago Alex fell ill and had to be left alone at the vet's overnight. He was so distressed when Irene went to leave the room he cried, "Come here, I love you, I'm sorry. Wanna go back."

At the Kewalo Basin Marine Mammal Laboratory at the University of Hawaii, Louis M. Herman has been studying the ability of two Bottle-nose dolphins to comprehend sentences. Phoenix has been taught acoustic language. Computer-generated sounds are broadcast into the tank through an underwater speaker. The second dolphin, Akeakamah, is taught in gestural language. Words are expressed by using the trainer's arm and hands as in sign language. Both dolphins have been shown to comprehend sentence form and meaning way above chance.

Testing the Case for Rights

But just how far are we going to take this concept of giving animals legal rights? The animal rights law advocates are not claiming that every species deserves the same legal rights. The case must be examined for different species.

"There is a lot we need to learn," says Wise, "but my aim is to teach a generation of lawyers who can begin test cases in a few years when we are prepared to win them." He has already litigated to improve the lot of some animals but as the law stands with animals regarded as property it is hard to get a good result. One encouraging case, though, was that of Rainbow the dolphin. In 1991 to 92, Rainbow was being held by an aquarium who wanted to give him to the navy for dangerous projects. Wise sued the government and the aquarium, and the dolphin was allowed to stay where he was.

This case illustrates the problem of animals being regarded as property. If something is deemed property, it is hard to stop the owner doing what they want even if—as in Rainbow's case—it could have endangered his life.

In October [2000] New Zealand's Parliament achieved a world-first when it passed legislation prohibiting the use of all great apes in research, testing or teaching unless such use is "in the best interests of the non-human hominid" or "in the interests of the species to which it belongs." That primates should have a right to life, liberty and freedom from torture on the basis of scientific evidence that they not only share our genes but also demonstrate self-awareness and cognitive, emotional and social capacities, is an argument rapidly gaining ground.

Maybe if you'd had a conversation with an ape, you'd agree.

JUSTIFYING THE ANIMAL RIGHTS POSITION

Gary L. Francione

Gary L. Francione is professor of law and Nicholas de B. Katzenbach Scholar of Law and Philosophy at Rutgers University Law School in Newark, New Jersey. A staunch advocate of animal rights, Francione is frequently confronted with questions about his beliefs. In this excerpt from his book *Animal Rights: Your Child or the Dog?* Francione addresses some of the questions he usually hears. His answers place an emphasis on the importance of considering animals as morally equal to humans. Francione argues that animals have sentience—that is, they can feel, suffer, and make choices. He stresses that because of this, animals should be held in equal regard to humans. Francione's responses to these common questions argue for moral equivalence for all sentient creatures.

Question: If you are in favor of abolishing the use of animals as human resources, don't you care more about animals than you do about those humans with illnesses who might possibly be cured through animal research?

Answer: No, of course not. This question is logically and morally indistinguishable from that which asks whether those who advocated the abolition of human slavery cared less about the well-being of southerners who faced economic ruin if slavery were abolished than they did about the slaves.

The issue is not whom we care about or value most; the question is whether it is morally justifiable to treat sentient beings—human or nonhuman—as commodities or exclusively as means to the ends of others. For example, we generally do not think that we should use any humans as unconsenting subjects in biomedical experiments, even though we would get much better data about human illness if we used humans rather than animals in experiments. After all, the application to the human context of data from animal experiments—assuming that the animal data are relevant at all—requires often difficult and always imprecise extrapolation. We could avoid these difficulties by using humans, which would eliminate the need for extrapolation. But we do not do so because even though we may disagree about many moral issues, most of us are in agreement that the use of humans as unwilling experimental subjects is ruled out as an option from the beginning. No one suggests that we care more about those we are unwilling to use as experimental subjects than we do about the others who would benefit from that use. . . .

Question: Where do you draw the line on who can have rights? Do insects have rights?

Answer: I draw the line at sentience because, as I have argued, sentient beings have interests and the possession of interests is the necessary and sufficient condition for membership in the moral community. Are insects sentient? Are they conscious beings with minds that experience pain and pleasure? I do not know. But the fact that I do not know exactly where to draw the line, or perhaps find drawing the line difficult, does not relieve me of the obligation to draw the line somewhere or allow me to use animals as I please. Although I may not know whether insects are sentient, I do know that cows, pigs, chickens, chimpanzees, horses, deer, dogs, cats, and mice are sentient. Indeed, it is now widely accepted that fish are sentient. So the fact that I do not know on what side of the line to place insects does not relieve me of my moral obligation to the animals whom I do know are sentient.

As a general matter, this question is intended to demonstrate that if we do not know where to draw the line in a matter of morality, or if line drawing is difficult, then we ought not to draw the line anywhere. This form of reasoning is invalid. Consider the following example. There is a great deal of disagreement about the scope and extent of human rights. Some people argue that health care and education are fundamental rights that a civilized government should provide to everyone; some people argue that health care and education are commodities like any other, not the subject of rights, and that people ought to pay for them. But we would, I suspect, all agree that whatever our disagreements about human rights—however unsure we are of where to draw the line—we most certainly agree, for instance, that genocide is morally wrong. We do not say that it is morally acceptable to kill off entire populations because we may disagree over whether humans are entitled to health care. Similarly, our uncertainty or disagreement regarding the sentience of ants is no license to ignore the interests of chimpanzees, cows, pigs, chickens, and other animals whom we do know are sentient. . . .

Question: Isn't taking advantage of medications or procedures developed through the use of animals inconsistent with taking an animal rights position?

Answer: No, it is not. Those who support animal exploitation often argue that accepting the "benefits" of animal use is inconsistent with criticizing the use of animals.

This position, of course, makes no sense. Most of us are opposed to racial discrimination, and yet we live in a society in which white middle-class people enjoy the benefits of past racial discrimination; that is, the majority enjoys a standard of living that it would not have had there been a nondiscriminatory, equitable distribution of resources, including educational and job opportunities. Many of us support measures, such as affirmative action, that are intended to correct past discrimination. But those who oppose racial discrimination are not obligated to leave the United States or to commit suicide because we cannot avoid the fact that white people are beneficiaries of past discrimination against people of color.

Consider another example: assume that we find that the local water company employs child labor and we object to child labor. Are we obligated to die of dehydration because the water company has chosen to violate the rights of children? No, of course not. We would be obligated to support the abolition of this use of children, but we would not be obligated to die. Similarly, we should join together collectively and demand an end to animal exploitation, but we are not obligated to accept animal exploitation or forego any benefits that it may provide.

We certainly could develop drugs and surgical procedures without the use of animals, and many would prefer we do so. Those who object to animal use for these purposes, however, have no control as individuals over government regulations or corporate policies concerning animals. To say that they cannot consistently criticize the actions of government or industry while they derive benefits from these actions, over which they have no control, is absurd as a matter of logic. And as a matter of political ideology, it is a most disturbing endorsement of unquestioned obeisance to the policies of the corporate state. Indeed, the notion that we must either embrace animal exploitation or reject anything that involves animal use is eerily like the reactionary slogan "love it or leave it," uttered by the pseudo-patriots who criticized opponents of American involvement in the Vietnam War.

Moreover, humans have so commodified animals that it is virtually impossible to avoid animal exploitation completely. Animal byproducts are used in a wide variety of things, including the asphalt on roads and synthetic fabrics. But the impossibility of avoiding all contact with animal exploitation does not mean that we cannot avoid the most obvious and serious forms of exploitation. The individual who is not stranded in a lifeboat or on a mountaintop always has it within her power to avoid eating meat and dairy products, products that could not be produced without the use of animals, unlike drugs and medical procedures, which could be developed without animal testing.

Question: Is it likely that the pursuit of more "humane" animal treatment will eventually lead to the recognition that animals have the basic

*right not to be treated as things, and the consequent abolition of institution-
alized animal use?*

Answer: No, it is not likely. Anticruelty laws requiring the humane
treatment of animals have been popular in the United States and
Great Britain for well over a hundred years, and we are using more
animals in more horrific ways than ever before. Sure, there have been
some changes. In some places, like Britain, veal calves get more space
and some social interaction before they are slaughtered; in some
American states, the leghold trap is prohibited and animals used for
fur products are caught in "padded" traps or raised in small wire cages
before they are gassed or electrocuted. Under the federal Animal Welfare Act, primates are supposed to receive some psychological stimula-
tion while we use them in horrendous experiments in which we
infect them with diseases or try to ascertain how much radiation they
can endure before they become dysfunctional. Some practices, such as
animal fighting, have been outlawed, but, as I have argued, such prohibitions tell us more about class hierarchy and prejudice than they
do about our moral concern for animals. All in all, the changes we
have witnessed as the result of animal welfare laws are nothing more
than window dressing.

This should not surprise us. Anticruelty laws assume that animals
are the property of humans, and it is in this context that the supposed balance of human and animal interests occurs. But as we saw,
we cannot really balance the interests of property owners against their
property because property cannot have interests that are protectable
against the property owner. The humane treatment principle, as
applied through animal welfare laws, does nothing more than require
that the owners of animal property accord that level of care, and no
more, that is necessary to the particular purpose. If we are using animals in experiments, they should receive that level of care, and no
more, that is required to produce valid data. If we are using purpose-
bred animals to make fur coats, they should receive the level of care,
and no more, that is required to produce coats that are soft and shiny.
If we are raising animals for food, those animals should receive that
level of care, and no more, that is required to produce meat that can
be sold at a particular price level to meet a particular demand. If we
are using dogs to guard our property, we should provide the level of
care that is required to sustain the dog for that purpose. As long as we
give the dog the minimal food and water and shelter—a dead dog will
not serve the purpose—we can tie that dog on a three-foot leash and
we can beat him, even excessively, for "disciplinary" purposes.

We claim to acknowledge that the interest of animals in not suffering is morally significant, but our animal practices belie that claim. If
we are really to honor the moral interests of animals, then we must
abolish institutionalized animal exploitation and not merely regulate
animal use through animal welfare measures that assume the legiti-

macy of the status of animals as property. . . .

Question: Isn't the matter of whether animals ought to be accorded the basic right not to be treated as our resources a matter of opinion? What right does anyone have to say that another should not eat meat or other animal products, or how they should otherwise use or treat animals?

Answer: Animal rights are no more a matter of opinion than is any other moral matter. This question is logically and morally indistinguishable from asking whether the morality of human slavery is a matter of opinion. We have decided that slavery is morally reprehensible not as a matter of mere opinion, but because slavery treats humans exclusively as the resources of others and degrades humans to the status of things, thus depriving them of moral significance.

The notion that animal rights are a matter of opinion is directly related to the status of animals as human property; this question, like most others examined here, assumes the legitimacy of regarding animals as things that exist solely as means to human ends. Because we regard animals as our property, we believe that we have the right to value animals in the ways that we think appropriate. If, however, we are not morally justified in treating animals as our property, then whether we ought to eat meat or use animals in experiments or impose pain and suffering on them for sport or entertainment is no more a matter of opinion than is the moral status of human slavery.

Moreover, as long as animals are treated as property, then we will continue to think that what constitutes "humane" treatment for your animal property really *is* a matter of opinion because you get to decide how much your property is worth. Just as we have opinions about the value of other things that we own, we can have opinions about the value of our animal property. Although our valuation of our property may be too high or too low relative to its market value, this is not generally considered a moral question. So when Jane criticizes Simon because he beats his dog regularly in order to make sure that his dog is a vicious and effective guard dog, Simon is perfectly justified in responding to Jane that her valuation of his property is not a moral matter up for grabs, but a matter of his property rights.

On another level, this question relates to . . . the position that all morality is relative, a matter of convention or convenience or tradition, with no valid claim to objective truth. If this were the case, then the morality of genocide or human slavery or child molestation would be no more than matters of opinion. Although it is certainly true that moral propositions cannot be proved in the way that mathematical propositions can, this does not mean that "anything goes." Some moral views are supported by better reasons than others, and some moral views have a better "fit" with other views that we hold. The view that we can treat animals as things simply because we are human and they are not is speciesism pure and simple. The view that we ought not to treat animals as things is consistent with our general notion that animals have morally significant interests. We do not treat any humans exclusively as the resources of others; we have abolished the institution of human property. We have seen that there is no morally sound reason to treat animals differently for purposes of the one right not to be treated as a thing, and that the animal rights position does not mean that we cannot prefer the human over the animal in situations of true emergency or conflict where we have not manufactured that conflict in the first place by violating the principle of equal consideration. . . .

Question: Of course the amount of animal suffering incidental to our use of animals is horrendous, and we should not be using animals for "frivolous" purposes, such as entertainment, but how can you expect people to give up eating meat?

Answer: In many ways this is an appropriate question with which to conclude our discussion because the question itself reveals more about the history of the human/animal relationship than any theory, and it demonstrates our confusion about moral matters in general.

Many humans like to eat meat. They enjoy eating meat so much that they find it hard to be detached when they consider moral questions about animals. But moral analysis requires at the very least that we leave our obvious biases at the door. Animal agriculture is the most significant source of animal suffering in the world today, and there is absolutely no need for it. Indeed, animal agriculture has devastating environmental effects, and a growing number of health care professionals claim that meat and animal products are detrimental to human health. We could live without killing animals and could feed more of the world's humans—the beings we always claim to care about when we seek to justify animal exploitation—if we abandoned animal agriculture altogether.

The desire to eat meat has clouded some of the greatest minds in human history. Charles Darwin recognized that animals were not qualitatively different from humans and possessed many of the characteristics that were once thought to be uniquely human—but he continued to eat them. Jeremy Bentham argued that animals had morally significant interests because they could suffer, but he also continued to eat them.

Old habits die hard, but that does not mean they are morally justified. It is precisely in situations where both moral issues and strong personal preferences come into play that we should be most careful to think clearly. As the case of meat eating shows, however, sometimes our brute preferences determine our moral thinking, rather than the other way around. Many people have said to me, "Yes, I know it's morally wrong to eat meat, but I just love hamburgers."

Regrettably for those who like to eat meat, this is no argument, and a taste for meat in no way justifies the violation of a moral principle. Our conduct merely demonstrates that despite what we say about the moral significance of animal interests, we are willing to ignore those interests whenever we benefit from doing so—even when the benefit is nothing more than our pleasure or convenience.

If we take morality seriously, then we must confront what it dictates: if it is wrong for Simon to torture dogs for pleasure, then it is morally wrong for us to eat meat.

CHAPTER 2

SHOULD ANIMALS HAVE LEGAL RIGHTS?

Contemporary Issues
Companion

ANIMAL RIGHTS ACTIVISTS ARE GAINING LEGAL PROTECTIONS FOR ANIMALS

Jim Motavalli

Should animals have legal rights? According to Jim Motavalli, editor of *E Magazine*, an environmental publication, the movement to grant legal protections to animals is gaining force. Motavalli cites laws that have been enacted to protect animals in the United States since the 1600s and explains that legal experts are becoming more sympathetic to the cause of animal rights. In fact, Motavalli points out, influential law schools such as Harvard are now offering courses in animal rights. The author relates that gaining legal protections for animals is difficult, but voter initiatives and government legislation are leading to slow but recognizable progress towards the recognition of animals' legal standing.

Does a pig packed in a tiny factory cage waiting to be killed have any rights in America? Should it have? And what about the chimpanzee, which shares 99 percent of its active DNA with humans? Should anyone be allowed to "own" an animal with so many of our own attributes, including the ability to reason, use tools and respond to language? Isn't that like slavery?

The fight to give animals legal rights barely registers on the environmental agenda, but perhaps it should. This isn't simply an endless philosophical debate but a gathering global force with broad implications for our planet's future, including how we use our natural resources. If animals had rights, we probably couldn't continue to eat them, experiment on them with impunity or wear their skins on our backs. Our fundamental relationship would change.

But precisely because our way of life depends on exploiting them, animals don't really have any significant "rights" in America, although Congress passed the Humane Methods of Slaughter Act (which requires simply that animals be "rendered insensitive to pain" before being killed) in 1958 and the Animal Welfare Act (which sets

Jim Motavalli, "Rights from Wrongs," *E Magazine*, vol. 14, March/April 2003, p. 26. Copyright © 2003 by *E/The Environmental Magazine*. Subscription Department: PO Box 2047, Marion, OH 43306. Telephone: 815-734-1242 (Subscriptions are $20 per year). On the Internet: www.emagazine.com. E-mail: info@emagazine.com. Reproduced by permission.

limited standards for humane care but exempts small laboratory animals) in 1966. All states afford animals some small measure of protection through anti-cruelty laws, but these laws have nothing to say about an animal's "right" not to be slaughtered, or used for any number of human purposes.

In 2003, however, a new and growing movement is trying to afford some genuine legal rights for animals. Buoyed by a growing awareness about animal intelligence and capacities, the courts, state governments—and the general public in statewide referenda—are enacting and enforcing new legislation.

Animal rights are back on the agenda, at least partly due to the release of the book *Dominion* by an unlikely author, White House speechwriter Matthew Scully. The book might have gotten some attention even if the writer came from the ranks of known animal sympathizers, but the fact that Scully is a self-described conservative and a Bush insider got it widely reviewed and discussed. Scully describes the Animal Welfare Act as "a collection of hollow injunctions, broad loopholes and light penalties when there are any at all." Animals, writes Scully, are "a test of our character, of mankind's capacity for empathy and for decent, honorable conduct and faithful stewardship."

Protecting Animals

The stewardship concept has a long history. Legal prohibitions against cruelty to animals in the U.S. date back as far as the Massachusetts Bay Colony's 1641 "Bodies of Liberties" ("No man shall exercise any Tirranny or Crueltie towards any Bruite creature which are usuallie kept for man's use," it said). But the use of the phrase "man's use" is telling—the statutes have always been limited to preventing "unnecessary" or "unjustified" pain, which leaves the laws subject to broad differences in judicial interpretation. But killing animals for food, sport, clothing or for scientific research has almost always been upheld by the law.

In 37 states, cruelty to animals is now a felony, and four new laws were enacted in 2002. Concerned Floridians succeeded [in] November [2002] in passing a constitutional amendment on the inhumane treatment of "factory farm" pigs. Also before voters [that] November: a ban on cockfighting in Oklahoma (it passed), a plan to issue special license plates to pay for spaying and neutering of pets in Georgia (it also passed) and a ban on animal cruelty in Arkansas (it was defeated).

In Europe, Germany has amended its national constitution to protect "the natural foundations of life" for people and animals. In 1992, Switzerland acknowledged that animals were "beings" through a constitutional amendment. In 2000, the High Court of Kerala in India handed down an opinion that states, "It is not only our fundamental duty to show compassion to our animal friends, but also to recognize and protect their rights. . . . If humans are entitled to fundamental rights, why not animals?"

The Great Ape Project, founded in 1993 "to include the nonhuman great apes within the community of equals," giving them fundamental protections of life, liberty and bodily integrity, has won its first great victory in New Zealand, which in 1999 banned most experimentation on "non-human hominids." There are loopholes that allow for testing if it is "in the best interests of the non-human hominid."

Peter Singer, cofounder of the Great Ape Project, a professor at Princeton and a pioneer in animal rights philosophy, said that the New Zealand law "may be a small step forward for great apes, but it is nevertheless historic. It's the first time that a parliament has voted in favor of changing the status of a group of animals so dramatically that the animal cannot be treated as a research tool." There are more than 3,000 great apes in captivity around the world, and Singer called on "other national parliaments to take up the initiative and carry it further."

A History of Denial

It may seem silly to have to argue that animals feel pain, make decisions and experience desires, but some theorists posit that they don't. According to R.G. Frey, author of the 1980 book *Interests and Rights: The Case Against Animals*, they might experience some pleasant or unpleasant "sensations," but have no real preferences, wants or desires, lack memory and expectation and can't make any plans or intend anything.

"Some anthropologists say that animals were the very first private property," says Jim Mason, an attorney, animal advocate and author of the book *An Unnatural Order*. "Before the concept of money existed, they were a major measure of wealth. It's ironic given that long history that we're now talking about eliminating their property status."

Since the 17th century, when philosopher René Descartes argued that animals had no souls and could neither think nor suffer, a consensus has been emerging that nonhuman creatures actually function on a much higher plane than was previously believed.

While Western religions have denied that animals have souls, an *ABC News* poll in 2001 found that 43 percent of respondents disagreed (and 17 percent were undecided). Darwin's theory of evolution did much to advance the idea that animals were not mere automata. The anti-vivisection movement, which opposes the use of animals in medical experimentation, gathered force in Britain and the U.S. after the Civil War (and shared some of the rights concepts imbued in the abolitionist movement). The concepts of women's rights and, later, gay rights also advanced wider conceptions of legal protection. The emerging fact that higher animals share much of their DNA with humans was certainly influential.

Recent animal rights law cases have turned on such questions as the rights of students to opt out of dissecting frogs or cats, and the privacy rights to the medical records of animals in zoos. Such cities as Berkeley and West Hollywood in California, Boulder, Colorado and Amherst, Massachusetts have changed the legal definition of pet owners to "guardians," and the Los Angeles City Council is considering a similar move.

Peter Singer's influential book *Animal Liberation*, which has sold 500,000 copies, offered a philosophical argument on behalf of animals that has been extended by such philosophers as *The Case for Animal Rights* author Tom Regan, who argues that Singer did not go far enough. Regan says that all uses of animals for food and experiments should be legally enjoined.

The Case for Reform

Animals are attracting a high-profile group of sympathizers these days, particularly at America's law schools, 25 of which now offer courses in animal rights law (up from five in the mid-1990s). The Chimpanzee Collaboratory's Legal Committee hosted a symposium at Harvard Law School [in] September [2002] that featured such scholars of the law as Professor Alan Dershowitz (who opined that "rights grow out of wrongs"), Cass Sunstein of the University of Chicago (who argued that animals regarded as property can still have rights under the law, and that "our culture is much more interested in protecting animals than our laws are"), David Favre of Michigan State (who said that animals may "cross the river" to legal rights over the "stepping stones" of incremental change), and acclaimed primatologist Jane Goodall (who said that legal rights might prevent the poaching and habitat destruction that is threatening Africa's great apes with extinction).

Another influential voice arguing for legal protection for animals is attorney Steven Wise, a former Harvard animal rights lecturer, a speaker at the chimpanzee symposium and the author of *Rattling the Cage: Toward Legal Rights for Animals* and *Drawing the Line: Science and the Case for Animal Rights*. Wise makes what he calls the "liberty" argument. He says that some nonhuman animals, including great apes, have "a kind of autonomy that judges should easily recognize as sufficient for legal rights." He also makes an "equality" argument, pointing out that children born severely retarded and dependent are automatically granted full human rights, and that "the principle of equality requires us to give [the same rights] to a bonobo who has high levels of cognition and a great deal of mental complexity." Wise's work is in part based on new research that finds, for instance, that some parrots "are probably self-aware, can grasp abstractions, imitate and use a sophisticated proto-language." A report in *Science* magazine [in early 2003] offers new evidence that orangutans and other apes exhibit cultural behavior.

Wise believes that the body of common law at the heart of American jurisprudence is flexible and based on fundamental values of liberty and equality. "When judges look at the principles of why humans have basic rights, I think they'll see that at least some nonhuman animals are entitled to rights for the same reason," Wise said in an interview. "To deny that is to be involved in arbitrary decision-making, which the common law frowns upon."

Wise points out that judges may be conservative, but so are the arguments he's making. "These values already exist," he says. Wise adds that any judge's decision in favor of animals having rights is likely to be appealed, and that he's really aiming to be heard on the appellate and state supreme court levels. His chances to succeed in the high courts, he believes, are enhanced by changes in public values (including a growing awareness of primate intelligence) and new scientific findings.

A major problem in making progress with animal rights law is the question of "standing" in the courts. Animals cannot generally be plantiffs in lawsuits. But Wise argues that the law makes many exceptions already. "Most judges already know that under the law as it stands, membership in any species is not enough by itself to entitle any being to legal person-hood," he says. "It is the dignity that derives from the ability to wield what I call a 'realistic, or practical autonomy' that is sufficient. Once any one legal right is given to any one non-human animal, the legal inquiry for basic rights can begin to shift from the question of 'are you a human being?' to 'do you have the necessary realistic autonomy?' The best initial candidate species, I believe, are the great apes, particularly chimpanzees and bonobos."

Expanding Rights for Animals

Steve Ann Chambers, president of the Animal Legal Defense Fund (ALDF), points out that ships, municipalities, trusts and, increasingly, multinational corporations (through what Kalle Lasn describes in *Culture Jam* as "their own global charter of rights and freedoms, the Multinational Agreement on Investment"), have the standing in the courts that is denied to animals. "We need to expand legal rights beyond humans," she says, adding that the law as currently written refuses animals legal standing, but also makes it nearly impossible for human plaintiffs to have standing on their behalf. Chambers points to a case in which ALDF sued the U.S. Department of Agriculture (USDA) for three counts of violating the Animal Welfare Act. "The lower court upheld our contention that violations had occurred," she says, "but when the government appealed, assisted by the biomedical industry, the appellate court said that ALDF had no standing."

A similar case involving Barney, a 19-year-old chimpanzee held by the Long Island Game Farm, had a bittersweet resolution. ALDF sued the USDA on behalf of Marc Jurnove, a frequent visitor to Barney who was disturbed by his isolation and neglect (contrary to 1985 provisions of the Animal Welfare Act that call on the agency to protect primates' "psychological well-being"). U.S. District Judge Charles Richey agreed that Barney had been abused, and he chided the USDA for not creating enforceable statutes for roadside zoos.

The government appealed, and it was during that process, in 1996, that Barney apparently got tired of waiting for justice. He fled his cage when someone forgot to lock it, scaled a seven-foot fence and was promptly dispatched with a shotgun. Three years later, there was finally a ruling in Barney's case: The appeals court reversed Richey's order to create new regulations, but it upheld Jurnove's right to be involved in the case. According to Joyce Tischler, executive director of ALDF, Jurnove won what is called "aesthetic standing," similar to the right people have to sue their local park because they're upset by the poor conditions of a scenic overlook. "We're trying to create incremental changes in the law," she says.

For Regan, professor emeritus at North Carolina State University and author of the book, *Empty Cages: The Future of Animal Rights*, animals should have the right to bodily integrity, freedom to live their lives according to their own needs and the overall right to life. "Unfortunately," he says, "we haven't made any real progress in achieving standing for animals in the legal system. We're still in a situation where you have to argue that something constitutes cruelty, which is very hard to prove. If you look at how the Animal Welfare and Humane Slaughter Acts work and are applied, it's just a farce. It's reminiscent of previous decades, when women and blacks couldn't get court standing."

Paul Waldau, who teaches the Harvard course in animal rights law previously taught by Wise, quotes Will Rogers as saying, "People who love sausage and respect the law should never watch either being made," and he describes the latter as "an ugly process dominated by monied interests." One of the biggest barriers to reform for lab animals is the National Institutes of Health (NIH) which, he says, "is not about to lose its right to use animals like chimpanzees in research. The NIH is opposed to even the most reasonable improvements." Chimpanzee language pioneer Roger Fours says some arrogant scientists think of apes as "hairy test tubes."

A Difficult Case for Legal Rights

Will overturning centuries of human dominion over animals be difficult? Wise admits it will be, but he notes that slavery was as deeply ingrained in the collective consciousness and that while it took a civil war to dislodge it in the U.S., in England it was overturned not by swords but by well-honed fountain pens, in the courts.

Dominion author Matthew Scully agrees with Wise that the great apes may be the best starting place to establish legal rights for animals. "There's a certain logic to that," he says from the Bush White

House, where he resumed speechwriting duties [in] December [2002]. Scully believes that existing laws already enshrine some protections for pets. "The law has some contradictions in regarding your dog as property, but also allowing that same dog to be the victim of crimes, including felonies," he says. "The law places moral boundaries around that animal, and makes some moral claims around it. That limits your rights of property and defines you as the animal's guardian." Scully believes that the law may eventually set aside the entire concept of animals as property and replace it with the legal guardian status implied in the anti-cruelty statutes.

In reviewing Wise's book *Drawing the Line* for the *Wilson Quarterly*, Scully observed that dolphins can correctly press levers marked "yes" and "no" in response to questions about whether a ball is in their tank, an African gray parrot named Alex can correctly identify objects, shapes, colors and quantities of up to six, and elephants are resourceful problem solvers. "What would legal personhood for, say, elephants amount to?" Scully asks. "Specific and well-enforced protections from the people who harm them—those engaged in the exotic wildlife trade, for example, or the vicious people who to this day still hunt elephants for trophies," he answers.

None of the available evidence adds up to a case for legal rights, say some scholars. Richard Posner, a federal judge and lecturer at the University of Chicago Law School, says, "It's just not feasible to equate animals with humans. There are too many differences." The biomedical community defends its work as simply necessary. "It is pretty easy to sit around a table and intellectualize about [Wise's] stuff and talk about what you're willing to give up," Frankie Trull of the Foundation for Biomedical Research told *The Daytona Beach News-Journal*, "until you or somebody you care about is hit with some terrible disease."

Report from the Field

Wayne Pacelle, the activist vice president of the Humane Society of the U.S. (HSUS), points out that between 1940 and 1990 only one statewide initiative protecting animals was approved by voters (it was a mourning dove hunting ban in South Dakota, later reversed). But since 1990 there have been 38 statewide ballot campaigns, with the pro-animal forces winning in 24 of them.

Pacelle, who was personally involved in 22 ballot campaigns (17 of which won), describes them as "demonstrating our political strength. They pay many dividends and serve as a training ground for activists." Pacelle is a much-hated figure in the hunting, trapping, game fighting and biomedical research communities, and his pronouncements are frequently posted on their websites. "Are you supporting the HSUS 'one step at a time' political agenda?" asks Americans for Medical Progress, which quotes Pacelle as envisioning the use of the initiative process for "companion animal issues and laboratory animal issues and other issues that are appropriate." The U.S. Sportsmen's Alliance posted an editorial accusing HSUS of "lies and deception" and Pacelle of "duping" Washington State voters.

Victories from Voters

It's not surprising that HSUS in general and Pacelle in particular inspire such ire, since the group's legal campaigns (run in coalition with many other organizations and local supporters) have been singularly successful. Since 1990, voters across America have approved measures, propositions and proposals to ban steel-jawed traps, prohibit airborne hunting of wolves, ban bear baiting, prohibit cockfighting, outlaw slaughter of horses for human consumption and prevent the expansion of greyhound racing tracks.

A major victory for animal groups [in] November [2002] was the 55-45 percent win on a Florida amendment to ban hog farm gestation crates, which confine pigs to two-foot by seven-foot cages while they're pregnant. The crates, animal supporters said, "inhibit practically every normal pig behavior," give rise to crippling foot and leg injuries and produce sores and infections. *When Elephants Weep* author Jeffrey Masson calls the Florida victory—one of the first to regulate a factory farming practice on cruelty grounds—as "pure good," adding, "I'm convinced that 500 years from now it will be illegal to kill any farm animal."

Why appeal directly to the voters? "Special interests often control key committees in the state legislature," says Pacelle, "and they can thwart the popular will, making it difficult to get bills passed. It's better to get it done with voter initiatives."

Animal Rights Legislation

HSUS and other groups have tried to get national legislation passed, but Congressional lobbying makes that nearly impossible. Most of the measures attached to a federal farm bill, which went after so-called "puppy mills" (which produce large numbers of dogs under poor conditions for a quick profit), opposed killing black bears for their gall bladders and attempted to legislate treatment of the "downer" cows that are handled by slaughterhouses, were eviscerated or turned into "studies" during House-Senate conferences. Only provisions combating cock and dog fighting were left.

HSUS and the Fund for Animals jointly sponsor a Humane Scorecard that rates politicians for their voting records on animal issues, a process that has led the group to endorse many Republicans, including Elizabeth Dole in her successful North Carolina Senate race. Republican U.S. Senators with pro-animal voting records include Robert Byrd of West Virginia, Bob Smith of New Hampshire (no longer in office) and John Warner in Virginia. Former veterinarian Wayne Allard, a Colorado Republican, has won the animal groups' favor for sponsoring legislation against cockfighting, though he's no friend of the environment. (His 2001 League of Conservation Voters score was 13 percent.) Major animal rights groups sponsor Humane USA, a political action committee whose fondness for Republicans helps explain its relatively rosy view of [the] November [2002] elections: 17 of its 23 Senate picks won, as well as 205 of its 214 House choices.

The Animal Legal Defense Fund has an equally successful record, and a 20-year history. Founded by Joyce Tischler in 1979 as Attorneys for Animal Rights, it held its first conference on animal rights law in 1980. Highlights of its two decades of fighting for animals include helping to block the importation of 71,500 rhesus monkeys from Bangladesh for use in research (1983), challenging veal farming in Massachusetts (1984), suing to prevent the Navy from using dolphins in defense work (1989), petitioning to have birds, rats and mice used in research protected by the Animal Welfare Act (1990), founding the first of what are now two dozen student-based college chapters (1993), working to prosecute purveyors of animal "crush" videos (1999) and suing to block wild horse roundups (2001). It is just starting work on a body of animal protection laws in China.

"Animal rights law is just now catching on," says ALDF President Steve Ann Chambers. "It's being taught in 25 to 30 law schools and is cited in legal textbooks. Mainstream law is no longer laughing at us." The University of Chicago's Cass Sunstein points out, "As more people in academia start discussing animal law and more law schools add courses on the subject, you're going to see more people practicing law who are committed to the well-being of animals. And that's going to have a huge impact."

The record is less successful on the legislative front, Chambers admits. "We'd like to see the interests of animals recognized in the legal system, with enforceable penalties," she says. There is no body of civil law that protects animals—as long as basic needs are cared for and there's no obvious cruelty, owners have the final say in how animals are treated. She cites the case of Moe, a 32-year-old chimpanzee who was kept for decades in a small cage in a Los Angeles suburb. ALDF tried unsuccessfully to get itself appointed as Moe's representative in the case (as a "Guardian ad Litem") when Los Angeles finally seized the chimp. (The story has a happy ending anyway: Moe ended up in a sanctuary.)

The law makes very slow progress. Chambers says 19 states now have laws recognizing animals as beneficiaries of estates; silly, perhaps, but a possible step in recognizing their standing in court.

The Naysayers

It's not only animal exploiters who have a problem with this incremental legal strategy. There are also detractors from the left, such as *Animal Equality* author Joan Dunayer, who criticizes Wise for not extending his rights concept to, among other things, honeybees. "We don't want a few nonhuman animals to be regarded as honorary humans. We want to get rid of humanness as the basis for rights," she says. And then there's Rutgers law professor Gary Francione, author of such books as *Rain Without Thunder: The Ideology of the Animal Rights Movement*. Until 1999, Francione directed the Rutgers-based Animal Rights Law Clinic, but he closed it down, claiming that "the American animal rights movement has collapsed" and become reformist, rather than radical.

Francione takes on nearly everyone. Though he once served as attorney for People for the Ethical Treatment of Animals (the most influential rights group today), he is now openly critical of the group for not being radical enough. He also has issues with ALDF, Steven Wise, Peter Singer, Wayne Pacelle, Tom Regan and most of the other animal activists cited in this story.

Fighting the System

Francione compares the laws governing animal ownership to those regulating slavery. "They're structurally similar in that they favor the owner's interests, as the slave laws did," he says. "If you examine anti-cruelty laws carefully, what you see is that the laws don't provide any more protection than is necessary for efficient exploitation of the animal. It's crazy to argue that we're ever going to get significant legal change from common law courts. If Congress passed a law making factory farming illegal, for instance, it would drive up the price of meat and people would be in the streets." The result, he says, is very small gains. He cites PETA'S celebration of the Burger King veggie burger, and Peter Singer's favorable comments about McDonald's decision to give battery hens more cage space. "Maybe Peter finds that thrilling; I do not," Francione says. "It is a clear indication that welfarist reform is useless."

One of Francione's more interesting complaints is against the legal reformers' willingness to work with Republicans who are otherwise terrible on progressive issues. "The only way we make sense is as a movement of the left," he says, "and that can't mean making alliances with anti-choice, pro-military politicians like Elizabeth Dole and Bob Smith." He also deplores PETA's "I'd Rather Go Naked Than Wear Fur" campaign as sexist, a view many other animal activists share.

Steve Ann Chambers has heard from Francione and other critics many times before and she thinks they offer no realistic solutions, since the American people are not likely to embrace the strict no-meat, no-dairy diet called veganism (Francione's choice and his basis for change) any time soon. "I find it more productive to work positively with what we have within the existing legal system, and build upon it," she says. "If we refuse to do anything about the problems that exist for animals until society has decided it is no longer proper to eat meat, well be waiting a long time."

ANIMALS SHOULD NOT HAVE LEGAL RIGHTS

Evan Gahr

According to Evan Gahr, giving legal rights to animals will erode the rights of humans. Gahr asserts that by treating animals as quasi-human, governments will ultimately not be able to enact laws that protect people, for example, from vicious breeds of dogs. Another consequence of legal rights for animals will be increased malpractice lawsuits aimed at veterinarians, the author points out, which will in turn increase the consumers' cost of obtaining medical care for their pets. Eventually, Gahr continues, animals may even be entitled to be plaintiffs in court cases. In the end, Gahr concludes, people's rights will be curtailed. Gahr is a senior fellow at the Hudson Institute, a public policy research organization.

The day before Thanksgiving 1998, an elderly man was brutally attacked near his Sudbury, Massachusetts, home.

No sooner were the predators apprehended, however, than the cultural elite offered its familiar litany of excuses and rationalizations for the crime. The perpetrators—who it turned out had terrorized neighborhood joggers and other residents—were merely wayward souls "socialized" to act like vicious animals. A high-priced defense attorney from an acclaimed law firm that has saved more than 200 vicious predators from death row was hired to defend the accused. Some two years later, the case continued.

Johnnie Cochran to the rescue? Actually, the animals in question are four-legged—a Rhodesian ridgeback and a pit bull. Their high-profile lawyer is Debra Slater-Wise, who works in a two person firm with her husband Steven Wise, legal architect of the burgeoning movement for "pet owner rights."

Most pets certainly are lovable and virtual "canine family members," to use PC parlance. But sorry, Fido, under the law you are nothing but an armchair. Since antiquity common law has regarded pets as virtually indistinguishable from other forms of property. That's why

Evan Gahr, "Fido Goes to Court," *The American Spectator*, December 2000–January 2001, p. 56. Copyright © 2001 by *The American Spectator*. Reproduced by permission.

veterinarians could only be sued for the monetary value of their "patients" and municipalities had wide latitude to seize and euthanize "vicious" dogs that maul or maim.

Over the last decade, however, a new breed of lawyers—some of them ideological soul mates of the animal rights movement, others salivating over a huge potential clientele (61 million pet-owning households, according to *USA Today*)—have convinced courts to accord animals quasi-human status. These legal beagles now win unprecedented damage awards from veterinarians. They file seemingly endless appeals to keep municipalities from exercising their long-established right to "put down" certifiably vicious dogs. At "doggie death row trials," they bring out "character witnesses" and animal behaviorists to testify on the dog's behalf. They convince courts to invalidate state and city ordinances that would bar or restrict ownership of pit bulls and other manifestly dangerous dogs.

The Cultural Context for Rights

Like so many other permutations of the rights revolution, this one has a cultural context.

Moral Relativism: There is no such thing as a bad dog; only a "good dog having a bad day," as Steven Wise described a "client" that mauled a Lincoln, Massachusetts resident. He used a similar argument to defend a German Shepherd that bit at least five people.

Psychobabble: Nicholas Dodman, author of *Dogs Behaving Badly*, explains that dogs that bite small children aren't necessarily vicious. They're probably afflicted with "interspecies dyslexia"—i.e., an inability to differentiate between genuine threats and humans.

Leniency for Criminals: Cracking down on crime only breeds more crime. American Dog Owners Association lawyer Marshall Tanick contends that laws against pit bulls could exacerbate the pooch's "antisocial" tendencies.

Understanding the Criminal and his "Rage": In New Jersey, authorities are required to consider whether a dog was provoked before it attacked. New York City has a similar measure on the books. That's right, "canine rage" is an acceptable legal defense.

Better Yale than Jail: "Pet incidents can only be reduced and the general public be better protected [through] education, not legislation," argues Adrianne Lefkowitz of the American Dog Owners Association. "It is commonly understood that socialization and understanding of behavior is the best way to not only make a community safer regarding the pets we share them with, but to make our pets better companions and less likely to wind up in the shelter."

Dysfunctional Family: "The breed of the dog is not the problem," the ASPCA contends, "the behavior of the owner is." Pit bulls, German Shepherds, and rottweilers get a bad rap because they tend to attract abusive owners, argues lawyer Karen Copeland.

Racism: In an argument redolent of feminist insistence that differences between men and women are merely "social constructs," the pet owner rights movement says it's impossible to make meaningful distinctions between species, or even tell them apart. When the American Kennel Club in 1998 listed certain dogs as "not good" for children, defenders of the ostracized species went ballistic. "To say that all these dogs are 'this' and these dogs are 'that,' that's racism, canine racism," yelped Carl Holder of the Dachshund Club of America.

Blame Society First: The Centers for Disease Control now opposes breed-specific legislation to curb dangerous dogs. "This entire issue isn't a problem of dog breeds as much as it is a social problem," CDC pediatrician and medical epidemiologist Julie Gilchrist recently proclaimed. Everywhere from Florida to Minneapolis, courts have struck down breed-specific bans.

In 1998, New York Mayor Rudy Giuliani pushed for legislation that would give the city's dangerous dog law more teeth. Faced with rabid opposition from the "pet owner rights" movement, the bill remains bottled up in the city council. The city can't even bar pit bulls from public housing residences, where they are reportedly the "pet" of choice among supposed drug dealers.

Liability Issues

Veterinarians now have more reason to worry when a patient bares his teeth. They could be held financially liable for botched dental work or otherwise sued for improper care. Once virtually immune to malpractice lawsuits, vets now face a litigation-happy bar—and judges who happily oblige. Pet owner rights advocates make no bones about this particular manifestation of judicial activism. In a 1995 New York University law review article, Debra Squires-Lee boasted that courts have "either refused to recognize animals as property or have used subterfuge"—by employing such dubious concepts as "intrinsic value" to award damages high above the pet's actual market value.

Onward and upward. The "nuisance value" of a lawsuit—or the amount insurers pay to settle without a lawsuit—jumped from $200 to $250 in the 1970's to more than $4,000, Animal Legal Defense Fund head Joyce Tischler happily notes. Damage awards, just $300 in the early 1990's, have skyrocketed to five figures. *USA Today* reports that [in] April [2000], a Costa Mesa, California, woman whose rottweiler cried continually after botched dental work won a $20,000 judgment against the vet.

Some folks may be hitting the jackpot in the litigation lottery, but the upshot could be much misery for pets and their owners. Vets worry the litigation explosion is likely to force a dramatic increase in their insurance rates, which today is dirt cheap at around $200. As the cost of increased premiums is passed along to pet owners, they may find they can no longer afford to take Fido to the vet, the American

Veterinarian Association contends. What next: universal health care for the animal kingdom?

Pet Owner Rights

Even well-behaved pets can end up in court, as some nasty custody battles attest. When lovebirds split, just who gets the cat or dog? And by what standard?

In California [in] December [2000], an Orange County court tackle[d] just these questions. At issue are two dogs, Guinness and Roxie, whose "parents," Brooks Brann and Patti Dalby, aren't getting along. [In 2000] Brann moved to Montana from the Newport, California, home he shared with Dalby. Now he wants Guinness to join him, in addition to $25,000 in punitive damages and compensation for loss of the dog's companionship. Dalby counters that Guinness should remain with his "sister" Roxie in their present home. (Both dogs are "fixed," lest anyone worry that an unplanned pregnancy would further complicate the issue.)

The case is a telling reminder why "pet owner rights" is despite its name ultimately inimical to property rights. If this were a mere property rights case, the dog's ostensible interest would be immaterial, no matter how much dogooders yelped on his behalf.

Animals as Plaintiffs

Animal Legal Defense Fund head Joyce Tischler recently proclaimed that "We're pushing the envelope until we can press a case in which the animal is the plaintiff." (Raise your right paw and repeat after me. . . .) But until every dog has his day in court, she and other lawyers will speak for them. "We are starting to file [friend of the court] briefs. We would like the court to realize that there's a third party that has an interest here."

Just who precisely is the third party? This is a classic bait and switch: Left-wing advocacy groups, aided and abetted by plaintiffs' lawyers, use the plight of the "disadvantaged" to increase their own power. The new rights supposedly accorded to our furry friends further fuel the litigation explosion—and further embolden left-wing advocates to infringe on individual property holders and government alike. (Back in 1991, Steven Wise made headlines with a lawsuit that ultimately kept the New England Aquarium from turning over a dolphin to the U.S. Navy for research purposes.)

Soon enough it's the pet owners' rights that are being curtailed. The Animal Legal Defense Fund, which boasts a network of some 600 lawyers nationwide, is supporting legislation in Pennsylvania that would make it a crime to leave a pet alone in the car while the owner is out of sight. [Between 1996 and 2000] 20 states have made the abuse of domestic animals a felony.

Is it still okay to have Fido fetch your slippers? Perhaps not for much longer. Steven Wise & Co. are determined to manufacture new rights for "non-human animals." In his new book *Rattling the Cage*, Wise, an adjunct professor at Harvard Law School, says his goal is to "break down" the legal "wall of separation" between animal and man. Sounds absurd. But not long ago, so did federal protection for discrimination against "fat people," and prohibitions against "lookism." If present trends continue, anyone who thinks Wise and his soulmates are barking up the wrong tree is in for some rude surprises.

ANIMALS DESERVE LEGAL RIGHTS

Author and law professor Steven M. Wise asserts that animals are entitled to the same legal rights as humans. In the past, Wise explains, animals were not considered to have the same autonomy as humans, and autonomy was a necessary precondition for having rights. However, Wise asserts that autonomy is not necessary for basic legal rights—human babies, for example, do not act autonomously, yet they have rights. Another argument against giving animals legal rights is that they are not capable of thinking the way humans do—that is, they cannot think rationally or experience emotions. On the contrary, Wise relates, evidence has accumulated that some animals, particularly nonhuman primates, have extraordinary mental capabilities. This new knowledge should lead society to grant animals the rights they deserve, Wise concludes. Wise teaches law at several universities.

For centuries, the right to have everything that makes existence worthwhile—like freedom, safety from torture, and even life itself—has turned on whether the law classifies one as a person or a thing. Although some Jews once belonged to Pharaoh, Syrians to Nero, and African-Americans to George Washington, now every human is a person in the eyes of the law.

All nonhuman animals, on the other hand, are things with no rights. The law ignores them unless a person decides to do something to them, and then, in most cases, nothing can be done to help them. According to statistics collected annually by the Department of Agriculture, in the United States this year, tens of millions of animals are likely to be killed, sometimes painfully, during biomedical research; 10 billion more will be raised in factories so crowded that they're unable to turn around, and then killed for food. The U.S. Fish and Wildlife Service and allied state agencies report that hundreds of millions will be shot by hunters or exploited in rodeos, circuses, and roadside zoos. And all of that is perfectly legal.

What accounts for the legal personhood of all of us and the legal thinghood of all of them? Judeo-Christian theologians sometimes

Steven M. Wise, "Why Animals Deserve Legal Status," *Higher Education*, vol. 47, February 2, 2001, p. B13. Copyright © 2001 by *The Chronicle of Higher Education*. Reproduced by permission of the author.

argue that humans are made in the image of God. But that argument has been leaking since Gratian, the 12th-century Benedictine monk who is considered the father of canon law, made the same claim just for men in his Decretum. Few, if any, philosophers or judges today would argue that being human, all by itself, is sufficient for legal rights. There must be something about us that entitles us to rights.

Defining Autonomy

Philosophers have proffered many criteria as sufficient, including sentience, a sense of justice, the possession of language or morality, and having a rational plan for one's life. Among legal thinkers, the most important is autonomy, also known as self-determination or volition. Things don't act autonomously. Persons do.

Notice that I said that autonomy is "sufficient" for basic legal rights; it obviously isn't necessary. We don't eat or vivisect human babies born without brains, who are so lacking in sentience that they are operated on without anesthesia.

But autonomy is tough to define. Kant thought that autonomous beings always act rationally. Anyone who can't do that can justly be treated as a thing. Kant must have had extraordinary friends and relatives. Not being a full-time academic, I don't know anyone who always acts rationally.

Most philosophers, and just about every judge, reject Kant's rigorous conception of autonomy, for they can easily imagine a human who lacks it, but can still walk about making decisions. Instead, some of them think that a being can be autonomous—at least to some degree—if she has preferences and the ability to act to satisfy them. Others would say she is autonomous if she can cope with changed circumstances. Still others, if she can make choices, even if she can't evaluate their merits very well. Or if she has desires and beliefs and can make at least some sound and appropriate inferences from them.

As things, nonhuman animals have been invisible to civil law since its inception. "All law," said the Roman jurist Hermogenianus, "was established for men's sake." And why not? Everything else was.

Unfortunately for animals, many people have believed that they were put on earth for human use and lack autonomy. Aristotle granted them a few mental abilities: They could perceive and act on impulse. Many Stoics, however, denied them the capacities to perceive, conceive, reason, remember, believe, even experience. Animals knew nothing of the past and could not imagine a future. Nor could they desire, know good, or learn from experience.

Extraordinary Minds

For decades, though, evidence has been accumulating that at least some nonhuman animals have extraordinary minds. [In 2000,] 7-year-old Kanzi—a bonobo who works with Sue Savage-Rumbaugh, a biologist at Georgia State University—drubbed a human 2-year-old, named Alia, in a series of language-comprehension tests. In the tests, both human and bonobo had to struggle, as we all do, with trying to make sense of the mind of a speaker. When Kanzi was asked to "put some water on the vacuum cleaner," he gulped water from a glass, marched to the vacuum cleaner, and dribbled the water over it. Told to "feed your ball some tomato," he could see no ball before him. So he picked up a spongy toy Halloween pumpkin and pretended to shove a tomato into its mouth. When asked to go to the refrigerator and get an orange, Kanzi immediately complied; Alia didn't have a clue what to do.

In the 40 years since Jane Goodall arrived at Gombe, she and others have shown that apes have most, if not all, of the emotions that we do. They are probably self-conscious; many of them can recognize themselves in a mirror. They use insight, not just trial and error, to solve problems. They form complex mental representations, including mental maps of the area where they live. They understand cause and effect. They act intentionally. They compare objects, and relationships between objects. They count. They use tools—they even make tools. Given the appropriate opportunity and motivation, they have been known to teach, deceive, and empathize with others. They can figure out what others see and know, abilities that human children don't develop until the ages of 3 to 5. They create cultural traditions that they pass on to their descendants. They flourish in rough-and-tumble societies so intensely political that they have been dubbed Machiavellian, and in which they form coalitions to limit the power of alpha males.

Twenty-first-century law should be based on twenty-first-century knowledge. Once the law assumed that witches existed and that mute people lacked intelligence. Now it is illegal to burn someone for witchcraft, and the mute have the same rights as anyone else.

Today we know that apes, and perhaps other nonhuman animals, are not what we thought they were in the prescientific age when the law declared them things. Now we know that they have what it takes for basic legal rights. The next step is obvious.

CHAPTER 3

IS USING ANIMALS FOR FOOD ETHICAL?

Contemporary Issues
Companion

FACTORY FARMING IS UNETHICAL

Bernard E. Rollin

According to philosophy professor Bernard E. Rollin, industrialized agriculture does not treat animals with dignity and respect.
Christian and Jewish teachings, Rollin writes, stress that human
dominion over animals includes a moral imperative to safeguard
animal life. The practice of confinement agriculture—in which
domestic animals such as pigs are kept in tiny crates too small for
them to move or lie down comfortably—harms animals in the service of large-scale, efficient production processes, Rollin explains.
He argues that a return to the ethical methods of animal farming
would lessen the suffering of domestic animals without endanger-
ing the world's food supply. Rollin is a professor of philosophy,
physiology, and biophysics at Colorado State University.

A young man was working for a company that operated a large, total-
confinement swine farm. One day he detected symptoms of a disease
among some of the feeder pigs. As a teen, he had raised pigs himself
and shown them in competition, so he knew how to treat the animals. But the company's policy was to kill any diseased animals with
a blow to the head—the profit margin was considered too low to
allow for treatment of individual animals. So the employee decided to
come in on his own time, with his own medicine, and cured the animals. The management's response was to fire him on the spot for violating company policy. Soon the young man left agriculture for good:
he was weary of the conflict between what he was told to do and how
he believed he should be treating the animals.

Consider a sow that is being used to breed pigs for food. The over-
whelming majority of today's swine are raised in severe confinement.
If the "farmer" follows the recommendations of the National Pork
Producers, the sow will spend virtually all of her productive life (until
she is killed) in a gestation crate 2½ feet wide (and sometimes 2 feet)
by 7 feet long by 3 feet high. This concrete and barred cage is often
too small for the 500- to 600-pound animal, which cannot lie down
or turn around. Feet that are designed for soft loam are forced to carry
hundreds of pounds of weight on slotted concrete. This causes severe

Bernard E. Rollin, "Farm Factories," *Christian Century*, vol. 118, December 19–26,
2001, p. 26. Copyright © 2001 by the Christian Century Foundation. Reproduced
by permission.

foot and leg problems. Unable to perform any of her natural behaviors, the sow goes mad and exhibits compulsive, neurotic "stereotypical" behaviors such as bar-biting and purposeless chewing. When she is ready to birth her piglets, she is moved into a farrowing crate that has a creep rail so that the piglets can crawl under it and avoid being crushed by the confined sow.

Under other conditions, pigs reveal that they are highly intelligent and behaviorally complex animals. Researchers at the University of Edinburgh created a "pig park" that approximates the habitat of wild swine. Domestic pigs, usually raised in confinement, were let loose in this facility and their behavior observed. In this environment, the sows covered almost a mile in foraging, and, in keeping with their reputation as clean animals, they built carefully constructed nests on a hillside so that urine and feces ran downhill. They took turns minding each other's piglets so that each sow could forage. All of this natural behavior is inexpressible in confinement.

The Origins of Factory Farming

Factory farming, or confinement-based industrialized agriculture, has been an established feature in North America and Europe since its introduction at the end of World War II. Agricultural scientists were concerned about supplying Americans with sufficient food. After the Dust Bowl and the Great Depression, many people had left farming. Cities and suburbs were beginning to encroach on agricultural lands, and scientists saw that the amount of land available for food production would soon diminish significantly. Farm people who had left the farm for foreign countries and urban centers during the war were reluctant to go back. "How you gonna keep 'em down on the farm now that they've seen Paree?" a song of the '40s asked. Having experienced the specter of starvation during the Great Depression, the American consumer was afraid that there would not be enough food.

At the same time, a variety of technologies relevant to agriculture were emerging, and American society began to accept the idea of technologically based economies of scale. Animal agriculture begin to industrialize. This was a major departure from traditional agriculture and its core values. Agriculture as a way of life, and agriculture as a practice of husbandry, were replaced by agriculture as an industry with values of efficiency and productivity. Thus the problems we see in confinement agriculture are not the result of cruelty or insensitivity, but the unanticipated by-product of changes in the nature of agriculture. Confinement-based agriculture contradicts basic biblical ethical teaching about animals. Yet despite the real problems in these farm factories, few Jewish and Christian leaders, theologians or ethicists have come forward to raise moral questions about them or the practices characteristic of this industry.

The Old Testament forbids the deliberate, willful, sadistic, deviant, purposeless, intentional and unnecessary infliction of pain and suffering on animals, or outrageous neglect of them (failing to provide food and water). Biblical edicts against cruelty helped Western societies reach a social consensus on animal treatment and develop effective laws. The Massachusetts Bay colony, for example, was the first to prohibit animal cruelty, and similar laws exist today in all Western societies.

The Ethics of Anticruelty

The anticruelty ethic served two purposes: it articulated concern about animal suffering caused by deviant and purposeless human actions, and it identified sadists and psychopaths who abuse animals before sometimes "graduating" to the abuse of humans. Recent research has confirmed this correlation. Many serial killers have histories of animal abuse, as do some of the teens who have shot classmates.

Biblical sources deliver a clear mandate to avoid acts of deliberate cruelty to animals. We humans are obliged, for example, to help "raise to its feet an animal that is down even if it belongs to [our] enemy" (Exod. 23:12 and Deut. 22:4). We are urged not to plow an ox and an ass together because of the hardship to the weaker animal (Deut. 22:10), and to rest the animals on the sabbath when we rest (Exod. 20:10 and Exod. 23:12). Deuteronomy 25:4 forbids the muzzling of an ox when it is being used to thresh grain, for that would cause it major suffering—the animal could not partake of its favorite food, and allowing it to graze would cost the farmer virtually nothing (also in 1 Cor. 9:9 and 1 Tim. 5:18). We are to save "a son or an ox" that has fallen into a well even if we must violate the sabbath (Luke 14:5), and to avoid killing an ox because that would be like killing a man (Isa. 66:3).

Other passages encourage humans to develop a character that finds cruelty abhorrent. We are to foster compassion as a virtue, and prevent insensitivity to animal suffering. The injunction against "boiling a kid in its mother's milk" (Exod. 23:19; Exod. 34:26; Deut. 14:21) is supported by Leviticus 22: 26–33, which commands us not to take a very young animal from its mother, and not to slaughter an animal along with its young. The strange story of Balaam and his ass counsels against losing one's temper and beating an animal (Num. 22) and Psalm 145 tells us that God's mercy extends over all creatures. Surely humans are being directed to follow that model.

The Shift to Industrialized Agriculture

As one of my colleagues put it, "The worst thing that ever happened to my department is the name change from Animal Husbandry to Animal Science." The practice of husbandry is the key loss in the shift from traditional to industrialized agriculture. Farmers once put animals into the environment that the animals were biologically suited for, and then augmented their natural ability to survive and thrive by providing protection from predators, food during famine, water during drought, help in birthing, protection from weather extremes, etc. Any harm or suffering inflicted on the animal resulted in harm to the producer. All animal experiencing stress or pain, for example, is not as productive or reproductively successful as a happy animal. Thus proper care and treatment of animals becomes both an ethical and prudent requirement. The producer does well if and only if the animal does well. The result is good animal husbandry: a fair and mutually beneficial contract between humans and animals, with each better off because of the relationship. Psalm 23 describes this concept of care in a metaphor so powerful that it has become the vehicle for expressing God's ideal relationship to humans.

Unnatural Conditions

In husbandry agriculture, individual animal productivity is a good indicator of animal well-being; in industrial agriculture, this link between productivity and well-being is severed. When productivity as an economic metric is applied to the whole operation, the welfare of the individual animal is ignored. Husbandry agriculture "put square pegs in square holes and round pegs in round holes," extending individualized care in order to create as little friction as possible. Industrial agriculture, on the other hand, forces each animal to accept the same "technological sanders"—antibiotics (which keep down disease that would otherwise spread like wildfire in close surroundings), vaccines, bacterins, hormones, air handling systems and the rest of the armamentarium used to keep the animals from dying.

Furthermore, when crowding creates unnatural conditions and elicits unnatural behaviors such as tailbiting in pigs or similar acts of cannibalism in poultry, the solution is to cut off the tail (without anesthetics) or debeak the chicken, which can cause lifelong pain.

There are four sources of suffering in these conditions:
• violation of the animals' basic needs and nature;
• lack of attention to individual animals;
• mutilation of animals to fit unnatural environments;
• an increase in diseases and other problems caused by conditions in confinement operations.

A few years ago, while visiting with some Colorado ranchers, I observed an example of animal husbandry that contrasts sharply with the experience described at the beginning of this article. That year, the ranchers had seen many of their calves afflicted with scours, a diarrheal disease. Every rancher I met had spent more money on treating the disease than was economically justified by the calves' market value. When I asked these men why they were being "economically irrational," they were adamant in their responses: "It's part of my bargain with the animal." "It's part of caring for them." This same ethical outlook leads ranchers to sit up all night with sick, marginal calves, sometimes for days in a row. If they were strictly guided by economics, these people would hardly be valuing their time at 50 cents per hour—including their sleep time.

Yet industrialized swine production thrives while western cattle ranchers, the last large group of practitioners of husbandry agriculture, are an endangered species.

Confinement Agriculture Is Immoral

Confinement agriculture violates other core biblical ethical principles. It is clear that the biblical granting of "dominion" over the earth to humans means responsible stewardship, not the looting and pillaging of nature. Given that the Bible was addressed to an agrarian people, this is only common sense, and absolutely essential to preserving what we call "sustainability."

Husbandry agriculture was by its very nature sustainable, unlike industrialized animal agriculture. To follow up on our swine example: When pigs (or cattle) are raised on pasture, manure becomes a benefit, since it fertilizes pasture, and pasture is of value in providing forage for animals. In industrial animal agriculture, there is little reason to maintain pasture. Instead, farmers till for grain production, thereby encouraging increased soil erosion. At the same time, manure becomes a problem, both in terms of disposal and because it leaches into the water table. Similarly, air quality in confinement operations is often a threat to both workers and animals, and animal odors drive down real property value for miles around these operations.

Another morally questionable aspect of confinement agriculture is the destruction of small farms and local communities. Because of industrialization and economy of scale, small husbandry-based producers cannot compete with animal factories. In the broiler industry, farmers who wish to survive become serfs to large operators because they cannot compete on their own. In large confinement swine operations, where the system rather than the labor force, is primarily migratory or immigrant workers hired because they are cheap, not because they possess knowledge of or concern for the animals. And those raised in a culture of husbandry, as our earlier story revealed, find it intolerable to work in the industrialized operations.

The power of confinement agriculture to pollute the earth, degrade community and destroy small, independent farmers should convince us that this type of agriculture is incompatible with biblical ethics. Furthermore, we should fear domination of the food supply by these corporate entities.

Returning to Husbandry

It is not necessary to raise animals this way, as history reminds us. In 1988 Sweden banned high confinement agriculture; Britain and the EU ban sow confinement. If food is destined to cost more, so be it— Americans spend an average of only 11 percent of their income on food now, while they spent more than 50 percent on food at the turn of the century. We are wrong to ignore the hidden costs paid by animal welfare, the environment, food safety and rural communities and independent farmers, and we must now add those costs to the price of our food.

If we take biblical ethics seriously, we must condemn any type of agriculture that violates the principles of husbandry. John Travis reported the following comments made by the Vatican [in] December [2000]:

> Human dominion over the natural world must not be taken as an unqualified license to kill or inflict suffering on animals. . . . The cramped and cruel methods used in the modern food industry, for example, may cross the line of morally acceptable treatment of animals . . . Marie Hendrickz, official of the Congregation for the Doctrine of the Faith, said that in view of the growing popularity of animal rights movements, the church needs to ask itself to what extent Christ's dictum, "Do to others whatever you would have them do to you," can be applied to the animal world.

It is a radical mistake to treat animals merely as products, as objects with no intrinsic value. A demand for agriculture that practices the ancient and fair contract with domestic animals is not revolutionary but conservative. As Mahatma Gandhi said, a society must ultimately be morally judged by how it treats its weakest members. No members are more vulnerable and dependent than our society's domestic animals.

PETA Ignores Progress in the Food Industry

Amy Garber and James Peters

In the following selection, Amy Garber and James Peters explain that major restaurant chains such as McDonald's and Yum! Brands—which owns KFC, Pizza Hut, and Taco bell, among others—are implementing new regulations intended to improve the welfare of animals raised for food. Part of the motivation for reform, the authors write, stems from attacks by People for the Ethical Treatment of Animals (PETA), an animal rights group that has aggressively targeted fast food restaurants. However, Garber and Peters report, the restaurants claim that they are motivated more by their desire to see animals treated as humanely as possible than by the actions of PETA. Reforms such as independent audit programs designed to monitor compliance with animal welfare guidelines have insured that food animals are treated more humanely than ever before. Despite this progress, according to the authors, PETA's attacks on the industry continue. Garber and Peters write for *Nation's Restaurant News*.

In its latest campaign against a fast-food behemoth, PETA, or People for the Ethical Treatment of Animals, has employed such heavy-handed tactics as creating a Web site that is filled with graphic images of animal abuse and even throwing fake blood on a CEO.

And when those maneuvers failed to produce the desired effect, PETA paid a visit to the city where the quick-service company is based. Several of the group's activists walked door-to-door to visit the foodservice executives' neighbors in an effort to make its case for stepped-up animal-handling guidelines. PETA representatives also made stops at local churches and restaurants, passing out leaflets, stickers and other literature.

PETA makes no qualms about its relentless strategy of targeting major restaurant chains, explaining that such harsh tactics—which are intended to grab headlines and sway public opinion—bring about reform much more quickly than does lobbying on Capitol Hill.

"We can adversely affect the bottom line and stock price of public companies like McDonald's and KFC in a way that we don't have the money or the influence to change the USDA or the agriculture committees in the House and Senate," says Bruce Friedrich, PETA's vegan-outreach director. "They [members of Congress] tend to be unwilling to pass legislation that is not supported by animal agribusiness. That is why we have focused on corporations that have customers to lose."

After initially attacking burger chains to improve practices in cattle slaughterhouses, PETA focused on improving conditions for chickens. As a result, the group has targeted Yum! Brands Inc. and its KFC chain. Yum, based in Louisville, Ky., also owns Pizza Hut, Taco Bell, A&W, Long John Silver's and Pasta Bravo.

PETA's Influence

In 2003 a PETA activist threw fake blood on Yum's chairman and chief executive, David Novak, as he entered a co-branded KFC-A&W unit in Germany. While the group initially denied a role in the attack, it later claimed credit for it.

PETA mounted another attack in July [2003] when it slapped Yum and KFC with a lawsuit accusing the chicken chain of making misleading statements about its poultry welfare guidelines. After KFC removed from its Web sites information about its guidelines, PETA dropped the litigation, but it continue[d] its anti-KFC campaign with a "world week of action" Sept. 28 through Oct. 4 [2003]. During that week PETA, h[e]ld demonstrations at KFC outlets "in all 50 states and in dozens of countries around the world," Friedrich says.

Nonetheless, farm animal experts have given the foodservice industry widespread credit for recent animal welfare initiatives. But PETA's aggressive strategy has become more than a thorn in the side for many restaurateurs, causing the National Restaurant Association to compare the group's approach to "terrorism."

Steve Grover, the NRA's vice president of health, safety and regulatory affairs, explains that PETA has "picketed in front of and burned down restaurants in the past. They have thrown blood on people, and this is one of their milder tactics. I think they are a radical activist group, and I would say that they add little to responsibly addressing this issue [of animal welfare]. They simply are not a group that we feel like we can work with, and we have no intention of responding to them directly."

Although PETA's influence—or lack thereof—remains a hot topic of debate among restaurant operators, animal welfare experts seem to agree that the foodservice industry has made tremendous progress in its treatment of animals [between 1999 and 2003]. Reforms include updated slaughterhouse designs created to reduce animal anxiety, larger cages for egg-laying hens and much less use of cattle prods.

"I saw more changes in 1999 than I had seen previously in my whole 30-year career," says animal behaviorist Temple Grandin, who is an associate professor of animal science at Colorado State University.

Industry Reform

Grandin and many of her colleagues point to fast-food giant McDonald's Corp., the world's largest burger chain, as the industry's pioneer, becoming the first national chain to mandate supplier guidelines. All of its producers combined provide it with more than 2.5 billion pounds of chicken, beef and poultry annually.

In 2000 McDonald's required its egg suppliers to double the living space for each caged hen to a minimum of 72 square inches. At the time, McDonald's—which purchases some 2 billion eggs annually—also said it would no longer buy eggs from farmers who debeak chickens.

"McDonald's started setting standards with its suppliers, and then Wendy's and Burger King followed with programs of their own," says Grandin, who sits on several independent committees that oversee animal welfare for McDonald's and other large fast-food chains.

Grandin designed many of the livestock-handling facilities used in meat plants in North America and around the world, and she is considered an expert in devising ways to reduce animals' suffering and anxiety. She also developed an objective scoring system for assessing handling of cattle and pigs at meat plants that McDonald's and others have used for the basis of their audit programs.

The typical cost for upgrading facilities ranges from about $500 to $5,000 per plant, depending on its age, according to Grandin. She adds that a minority of facilities require upgrades costing $10,000 to $20,000, "but those are huge plants." She insists that most changes are simple and inexpensive, such as keeping litter off the floor, fixing broken equipment and installing better lighting.

"Animals aren't afraid of getting slaughtered, because they don't know what it is," Grandin explains. "But they are afraid of the dark and scared of shadows."

A System for Oversight

The foodservice industry's next major step is to integrate all of the chains' various guidelines under a single system called the Animal Welfare Audit Program developed by the National Council of Chain Restaurants, the Food Marketing Institute and the producer community. The program, which will provide standardized data regarding animal welfare at livestock production and slaughter facilities from turkey to pork, will be fully operational by fall [2003], according to NCCR president Terri Dort.

"Our effort was to put together a very sincere, real program that is transparent and that has accountability built into the audit," Dort explains. "Now these companies can stand up and say, 'We are committed to humane treatment of animals. End of story. Move on.'"

Many experts voice support for a consolidated audit program. Grandin explains that previously some facilities were being audited eight times a year because each restaurant chain was doing its own assessment. "Now the system will consolidate into two or three commercial audit companies," says Grandin, who adds that the best-run plants also have the highest food safety standards.

The guidelines are designed to be long-standing and the NCCR and FMI do not plan to revise them substantially in the future in the face of anticipated and ongoing criticism from such groups as PETA. Dort emphasizes that PETA's tactics are not only "threatening" but also "counterproductive," as the industry has been working diligently to put together the auditing-standards program, which has been "a major resource drain," Dort says. She adds, "The activists haven't backed off and let these companies do their work, and that's really a shame."

In fact, some observers are concerned that animal rights groups might be gaining too much clout and also are misleading the public in their goals. Wesley J. Smith, a lawyer, author and senior fellow at the Seattle-based Discovery Institute, a nonprofit public-policy think tank, says: "The agenda of animal liberationists isn't the humane, proper treatment of animals. It is the end of all use of animals. And restaurants and the food industry had better learn that."

Treating Animals Humanely

But PETA's Friedrich says the group has no hidden agenda and is not deceiving anyone.

"Animals are not here for humans to eat," he explains. "Our ultimate goal is that other animals are not used as means to an end. That is part of our mission statement. We are a vegan group. But we are also realistic, and where we have the public on our side, we are willing to take that. Where we would like to see empty cages, in the interim we would like to see bigger cages. If we can't get animals complete freedom immediately, at least we want to make it better for them."

But Smith insists that PETA goes too far in its attempts to accelerate change.

"There is a difference between persuasion and coercion, and I think PETA crosses that line. If the industry gives into coercion, then I think it looks for more of the same because it works," Smith says.

Although PETA has faced widespread criticism of its tactics as "tasteless" and "insensitive," the group has no plans to alter the way it pressures restaurateurs.

"We are not trying to win a popularity contest as much as we are trying to improve conditions for animals," Friedrich says.

Smith advises restaurant chains to defend themselves with a two-pronged response. "No. 1 is to make sure that your ducks are in a row and that the animals are being treated appropriately under the laws, and even, if necessary, a little better than the law requires.

"And then, second, assure yourself that you cannot compromise with these fascists," he continues. "The more you give into them, the more powerful they become, and the more emboldened they are to ask for more and more and more, because the industry could never treat animals humanely enough to satisfy these people."

Changes at McDonald's

McDonald's was PETA's first restaurant target, in the late 1990s, but the company insists that it began addressing the issue a decade earlier. It was Henry Spera, a leading farm animal advocate, who raised the chain's awareness of conditions at the producer level, according to Bob Langert, senior director of social responsibility for the Oak Brook, Ill.-based fast-food chain.

"We have always prided ourselves on social responsibility and leadership," Langert says. "We, as a company, wanted to do the right thing. It is a very important part of our heritage all the way back to [founder] Ray Kroc."

He adds, "But we always felt like we were not doing enough," explaining that the "magical moment" for McDonald's took place in the summer of 1997 when Grandin first presented her audit program.

"We saw it as something that had tremendous potential," Langert says. "We immediately said we loved her program, and we wanted to adopt it. [Grandin and Spera] are our real heroes that helped catalyze change within our company and with our suppliers. It took us from initial efforts in the early '90s to immediately integrating it into the way we do business."

McDonald's created an independent committee of experts, which includes Grandin, to help develop animal-handling guidelines. As a result, McDonald's suppliers now have upgraded their facilities and improved their management techniques, according to Langert.

He adds that McDonald's first implemented Grandin's scoring system in 1999, and in 2002 the company completed 500 audits worldwide for beef, poultry and pork processing plants.

While McDonald's officials credit Grandin and Spera, PETA insists its 1999 public campaign against the fast-food giant really started the wheels of change through an 11-month offensive that included 400 demonstrations with "provocative" ads and materials.

Bad Publicity

Friedrich says the turning point was the highly publicized 1997 verdict in the trial dubbed McLibel, which gained attention as the longest trial in British history, running for more than 300 days. The case stemmed from McDonald's suing of two London Greenpeace activists for defamation in a leaflet titled, "What's Wrong with McDonald's?"

Although the judge ultimately ruled in favor of McDonald's—

explaining that the activists had not demonstrated the truth of all their wide-ranging criticisms—the burger chain lost points on the public-relations front, Friedrich says. He explains that the British court determined that McDonald's was responsible for several practices that were legal but "cruel." Friedrich says PETA tried unsuccessfully to pressure McDonald's for two years before it launched its attack.

But Janice Swanson, a professor of animal science and industry at Kansas State University, says PETA "didn't get the ball rolling. They did a stupendous job of picking up the issue, but there were a lot of animal protection groups, government agencies and scientists working on this for a long time" before PETA began its campaign against McDonald's.

Swanson, who like Grandin sits on McDonald's and other independent industry committees, explains that a "whole host of factors converged" in the late 1990s. She describes two "watershed" moments: McDonald's announcing cage-space requirements for egg-laying hens in 2000, and, before that, restaurant executives starting to interact with suppliers. Swanson credits the foodservice industry with taking "a studied approach," relying on scientific data and experts.

She adds: "Restaurant chains had a lot of questions, and they were not about to disrupt the process in an abrupt manner. That is probably why they invited a mix of scientists and consumer organizations to get a full feel for the issue and find out what their responsibility was in all of this."

However, Swanson also voices frustration over the length of time it took suppliers to heed the call from animal agriculture experts. She says the last five years have been "rewarding and a little bit of a smack in the face to some degree. We are really happy with some of these changes, but we have been talking about these things for 15 years with producers. Getting action was something that the equation lacked."

She explains that the only federal legislation governing animal agriculture is the Humane Slaughter Act of 1956, which only covers cattle.

Taking the Initiative

"The care of farm animals has little federal regulation," says Swanson, who from 1987 to 1992 worked for the USDA in the Animal Welfare Information Center. But she points out that corporate initiatives have generated change faster than government regulation. She says the phenomena are being watched by experts as far away as Europe who are interested in watching restaurant chains take the initiative.

The most recent example is KFC, which [in 2003] adopted guidelines for its poultry suppliers. However, PETA has no plans to stop with its assaults against the chicken chain. Among other demands, the group wants KFC to mandate that its suppliers kill chickens with gas, prohibiting the practices of electric stunning or throat slitting.

Jonathan Blum, a senior vice president for Yum, says the fast-food giant is committed to the humane treatment of animals. "We have developed guidelines with the leading experts in the field," he says. "We have conducted both announced and unannounced visits of our suppliers to make sure they are in compliance with our guidelines. We take animal welfare seriously. We took a leadership role in our industry by establishing standardized guidelines for the entire poultry industry, including quick-service restaurants."

When it comes to PETA's recent assault on KFC and Yum, Blum responds: "I respect any individuals' rights to express freedom of speech. They are entitled to their opinion, even if we don't agree with them. In various instances they may have crossed the line. I don't have any comment on their publicity stunts."

But the Discovery Institute's Smith warns operators to resist caving in to PETA's demands. In a recent report he strongly advised such chains as KFC to "stand firmly against activists' intimidation whenever and however it occurs—even if it means that their presidents' homes are picketed, their stockholder meetings disrupted, and their executives' business suits stained by fake blood and pies in the face."

KFC TREATS CHICKENS INHUMANELY

People for the Ethical Treatment of Animals

People for the Ethical Treatment of Animals (PETA) is the world's largest animal rights organization. It opposes the use of animals for food, clothing, entertainment, and medical research. In the following selection, PETA focuses on the treatment of chickens by the fast-food chain KFC. According to PETA, these chickens are forced to live in cramped conditions, frequently suffer broken bones, and are boiled while conscious. While KFC has promised to improve its treatment of chickens, it has failed to do so. PETA offers several recommendations, including more living space and more humane methods of gathering and slaughtering the birds.

PETA is asking KFC to eliminate some of the worst abuses that chickens suffer on the factory farms and in the slaughterhouses of its suppliers, including live scalding, life-long crippling, and painful debeaking.

Chickens are inquisitive and interesting animals and are thought to be at least as intelligent as dogs or cats. When in natural surroundings, not on factory farms, they form friendships and social hierarchies, recognize one another, love their young, and enjoy a full life, dust bathing, making nests, roosting in trees, and more.

The more than 700 million chickens raised each year for KFC aren't able to do any of these things. They are crammed by the tens of thousands into sheds that stink of ammonia fumes from accumulated waste; they are given barely even room to move (each bird lives in the amount of space equivalent to a standard sheet of paper). They routinely suffer broken bones from being bred to be top heavy, from callous handling (workers roughly grab birds by their legs and stuff them into crates) and from being shackled upside down at slaughterhouses. Chickens are often still fully conscious as their throats are cut or when they are dumped into tanks of scalding hot water to remove their feathers. When they're killed, chickens are still babies, not yet two months old, out of a natural life span of 10-15 years.

In May of 2001, KFC's parent company, Yum Brands, Inc. assured PETA that it intended to "raise the bar" on animal welfare; yet, to date, KFC has done nothing to address some of the most egregious animal cruelty in the chicken industry. . . .

Chickens are probably the most abused animals on the face of the planet. They suffer any number of cruelties, including being left by the hundreds of thousands to starve to death, having their sensitive beaks seared off with hot blades, being crammed 11 birds to a tiny cage along with the decomposing corpses of other chickens, and dying in huge numbers from long journeys in extreme weather conditions. Basically, any and all abuse is allowable when it comes to chickens, who are, in fact, remarkable animals with distinct personalities and intelligence that, if allowed to develop, is as advanced as that of cats and dogs. Most importantly, they feel pain, just as we do.

PETA's Recommendations

The following is a basic outline of PETA's recommended animal-welfare program, followed by a review of the most egregious abuses inflicted on chickens who are raised for food.

• *Replace electrical stunning and throat slicing with contained-atmosphere stunning-to-kill.* Experts agree that contained-atmosphere stunning-to-kill causes less suffering for birds than KFC's present method of snapping chickens' legs into metal shackles and slicing their throats open, often while they are still conscious.

• *Install cameras in slaughterhouses to enforce humane standards.* Cameras should be installed at key points for animal handling, including unloading areas, the point of entry into the "stun" bath, the point of entry into the scalding tank, and places where chickens have their throats slit.

• *Switch to humane mechanized chicken gathering.* Studies have shown that when using manual methods, there are four times as many broken legs, more than eight times as much bruising, and increased stress.

• *Stop forcing rapid growth and feeding chickens drugs, and breed for health.* Breed leaner, healthier, less aggressive birds instead of breeding the biggest, fattest birds possible, and stop feeding chickens antibiotics and other drugs for nontherapeutic purposes.

• *Give chickens more living space.* Currently, bird fatality and injury rates are extremely high, based in part on the fact that the birds simply do not have enough space to survive. Experts agree that increased living space would decrease these problems.

• *Allow birds the opportunity to fulfill their natural desire for activity.* For example, provide the birds with whole green cabbages suspended in the air to peck at and eat. The cabbages stimulate healthy activity, dispel boredom, strengthen leg muscles, and provide nutrients without adding to the weight problems of these birds. Or include sheltered areas and perches in chicken houses, which would enhance the birds' living space, reducing their stress and aggression, and allow them to engage in some of their natural behaviors.

Overweight and Neglected

Chickens today are typically crammed by the tens of thousands into sheds, each chicken with less living space than a standard sheet of paper. Modern chickens' upper bodies grow six to seven times as fast as they used to—they are fed drugs and are genetically bred to grow so large and so quickly that their legs, lungs, and hearts often can't keep up. Many of these animals, whose lives are miserable from birth, suffer lung collapse, heart failure, and crippling leg deformities. On the farm cruelty is totally unregulated, extreme abuse goes completely unnoticed, as discovered when PETA went undercover on a turkey farm in Minnesota.

The breeding animals who "supply" the nation's 9 billion chickens have been called *Gallus neglectedus* ("neglected chickens") because their welfare is so often ignored entirely. These birds suffer from many of the same conditions forced on other chickens but suffer from them for a longer period of time. Additionally, the birds are forced to endure constant hunger from food deprivation and painful mutilations, including having their beaks seared off, their toes, spurs, and combs chopped off, and intranasal implants (plastic stick-like objects) inserted through male birds' nasal cavities so that the ends protrude horizontally to both sides of their faces, preventing the birds from reaching through the cage to eat the females' food.

Transport and Slaughter

Accidents are common in bird transport because "producers" know that they can sell the animals even after accidents. A few years ago, one driver testified that there had been four turkey truck accidents in one two-week period to meet the Thanksgiving demand. PETA documented one case where a truck turned over at 6 a.m., and the injured animals lay bruised and broken for six hours until another transport truck arrived to transport them to the slaughterhouse another five agonizing hours away. Humane handling standards must be required.

At slaughter, chickens are dumped from cages like so many rubber balls and then SNAPPED by their weakened and sometimes broken legs into metal shackles before their heads are passed through an electrically charged water bath that immobilizes them but often does not render them unconscious. The workers who hang the animals must work so quickly (assembly-line style) that animals are frequently injured. When the water "baths" are set below the level required to kill them, as they often are, the animals (unless they have died from stress and abuse before they're even shackled) are alive, conscious and bleeding to death after their throats are slit, and they enter the scalding tank (scalding hot water for feather removal) still conscious. Many of them flap about and thus miss both the immobilization bath and the automated and manual (human) neck-slicers and are still completely conscious when they are scalded to death.

Every time PETA and other animal groups go undercover on a factory farm or in a slaughterhouse, the level of abuse found is shocking—going even beyond the simple horrors that are routine. Unfortunately, we simply can't trust corporations or factory farmers to adhere to guidelines.

KFC must hire independent auditors and report audit progress on its Web site. KFC has hired Drs. Temple Grandin, Ian Duncan, and Joy Mench as consultants. They would make an excellent team to train and oversee independent auditors to implement the comprehensive program that PETA is proposing—which must include unannounced audits (verifiability) and a complete detail of the plan and progress (transparency).

PETA's Campaign Against KFC Reveals Its Anti-Meat Agenda

Jay Nordlinger

Jay Nordlinger is managing editor of *National Review*, a conservative journal. In this selection, he comments on the campaign being waged against fast food giant KFC by People for the Ethical Treatment of Animals, or PETA. Although PETA's campaign hinges on forcing KFC to employ more humane methods of raising and killing chickens, Nordlinger relates that their real goal is to eliminate the use of animals for any reason. Nordlinger argues that some of PETA's more moderate demands are reasonable, and KFC would be wise to adopt them if it meant better welfare for chickens. On the other hand, he reasons, PETA's more radical demands reveal the group's actual agenda of liberation for all animals. KFC should fight for its right to be in the chicken business, Nordlinger concludes.

Eaten at KFC lately? Well, shame on you—at least that's what PETA would say. PETA, of course, is People for the Ethical Treatment of Animals, the most notorious and influential animal-rights group in the country. And KFC is the chain once known, more amply, as Kentucky Fried Chicken. PETA has been on a fierce campaign against KFC, charging that the company treats chickens inhumanely, or at least allows its suppliers to. KFC, naturally, denies the charge. Who's right? And what is a fan of the Colonel, who is nevertheless a foe of animal abuse, to do?

Start with the fact that PETA does not or should not, inspire trust in normal people. Despite its beautiful name—who can be against the "ethical treatment of animals"?—PETA is an extremist group. They hold that meat-eating is "Holocaust on Your Plate." Sharks that maim and kill people have exacted "revenge." School lunches are "weapons of mass destruction"—that sort of thing.

PETA and Violence

More damnably, the group has clear ties to terrorism: For example, it has donated to the Earth Liberation Front, which is number one on

Jay Nordlinger, "PETA vs. KFC," *National Review*, December 22, 2003, p. 27. Copyright © 2003 by National Review, Inc., 215 Lexington Ave., New York, NY 10016. Reproduced by permission.

the FBI's domestic-terror list. In truth, PETA may, in its heart of hearts, see itself as somewhat soft: Ingrid Newkirk, the group's flamboyant and endlessly energetic president, has said, "If I had more guts, I'd light a match"—to research laboratories. And a statement by PETA's Bruce Friedrich is often cited by people who worry that the extent of PETA's radicalism is little known. Before a convention audience, he celebrated "blowing stuff up and smashing windows" and declared, "It would be great if all the fast-food outlets, slaughterhouses, these laboratories, and the banks who fund them exploded tomorrow. . . . Hallelujah to the people who are willing to do that!"

So, this is not a nice lil' anti-cruelty group. Violence aside, their aim is "total animal liberation," which is to say, no meat, no milk, no anything. They're against seeing-eye dogs for the blind, any and all medical research involving animals—the works. They can hardly serve as champions of those who want simply that their meat be raised and killed humanely.

But PETA has scored inarguable successes. It campaigned against McDonald's, Burger King, and Wendy's and ceased those campaigns when the companies undertook animal-welfare reforms. (The companies denied that the changes had anything to do with activist pressure, but then they probably would, wouldn't they?) PETA next turned its sights on KFC. Now, poor KFC gets its chickens from the same places everyone else does—Tyson's, Perdue, and the handful of other suppliers in this country. But KFC is a big name, and a big chain, so PETA went after the Colonel—the Colonel of Cruelty, they call him.

They put up a website, KentuckyFriedCruelty.com, which is a model of invective, camp, and gonzo aggression. PETA depicts the Colonel—kindly old Harlan Sanders—as a chicken-torturing maniac, complete with dripping knife. The first step in his "secret recipe" is, "Starve the parent birds constantly . . ." And you should check out the Kids' Corner, which, like the knife, features dripping Mood. "Chickens should be friends, not food!" PETA admonishes. The sermon concludes, "The best way to make sure that animals don't suffer is to stop eating them!"

Campaigning for Veganism

And there's the rub I mentioned: In the PETA view, the basic cruelty is the use of animals for food at all. These activists communicate a mixed message, as when they stand in front of KFC outlets holding banners that say "Go Vegan!" This must not cut much ice with those who, no matter what their concern for animal rights, aren't goin' vegan. And does Bruce Friedrich lose some effectiveness when, in letters to KFC urging more humane practices, he identifies himself as PETA's "Vegan Campaign Coordinator"? KFC can say—and does say—quite persuasively, Don't listen to these kooks. Their real aim is that you don't eat any chicken at all—or drink milk or own a cat. Come on!

PETA has been pulling anti-KFC stunts all over the country—nay, the world. In Paris, they had a "demo"—their word, a typical radical's word—in which they smeared fake blood and scared away customers. ("The protest blocked traffic on the busy Boulevard Sébastopol for more than two hours!") In Germany, an activist daubed the CEO of KFC's parent company with blood and feathers, causing Friedrich to remark, in characteristic fashion, "There is so much blood on this chicken-killer's hands, a little more on his business suit won't hurt."

PETA has enlisted a number of celebrities in its war against KFC, most prominently Pamela Anderson, the ex-Baywatch babe. She has done a provocative poster for the group, and has furnished—or so we're asked to believe—"Pamela's Festive Favorites," which are "animal-free" recipes (e.g., Best-Ever Green Bean Bake). And PETA put the arm on actor Jason Alexander, who was KFC's ad spokesman: They threatened to picket the show he was doing in L.A. (The Producers). He and KFC parted company, for reasons that are contested. What is not contested is that it's not good to cross PETA, unless you want a world of grief, unrelentingly.

The group campaigns with the zeal—and bullying passion—of fanatics. They harass KFC officials at home and at church (yes). They knock on neighbors' doors, imploring these folk to lobby the officials. At first, KFC tried cooperating with PETA, listening to its appeals and discussing reforms. But at some point, apparently, it determined that PETA is unappeasable.

Who Can You Believe?

In getting to the nitty-gritty, let's stipulate that chicken-raising and -killing is a dirty, unpretty business. It's almost a cliche to say that, if you've ever been on a chicken farm, you don't ever want to eat a chicken. PETA circulates a film of abuses at a chicken plant, and the film is virtually impossible to watch. KFC says that it is misleading, as the practices depicted are either obsolete or aberrant. Whom to believe? In many of these disputes, it comes down to he said, she said—and what common sense and intuition tell us. KFC has formed an "animal-welfare advisory panel," which sounds encouraging. PETA charges that the panel is a) stacked with corporate stooges and b) disregarded by the company anyway.

PETA says that its demands are relatively modest: that suppliers replace "electrical stunning and throat slicing" with "gas killing." (Talk about "Holocaust on Your Plate," but never mind.) That they institute "humane, mechanized chicken gathering," as opposed to "manual methods," which—contrary to what you might suppose—subject chickens to more injuries. That suppliers "Stop forcing rapid growth" and adopt as their primary concern the health of the chicken.

KFC's defenders say that some of PETA's demands are impossible, and meant to be impossible. And it would, indeed, seem too much to ask that suppliers breed "leaner, stronger" chickens, as PETA insists—restaurants, and customers, want their birds plump and juicy, unalterably. But some of the demands seem quite reasonable. When PETA is in moderate mode, it comes off as . . . well, non-crazy, and just. Why not "mechanized gathering" instead of the manual variety, particularly in an ever-mechanizing world? And why not "gas killing," as a gentler alternative to the other stuff? KFC itself says that it is looking into the feasibility of such a change.

KFC Should Defend Itself

I should note that the company does little to defend itself, and to help those who would be its friends. If you're a journalist—even from a conservative opinion magazine!—they won't talk to you. They will only fax you a two-sentence statement from a spokeswoman saying, in essence, Don't worry, be happy! KFC might show a little more self-confidence, and engage with its critics, when those critics seem engageable (i.e., non-crazy). In July [2003], PETA sued KFC for making false statements in its communications to the public. In September, it dropped the suit—because KFC dramatically altered its line. This was a black eye for the company, and a rather startling victory for PETA.

There are countless details in this case, and many twists and turns to the story. I will boil down for you, if you like, what I have concluded about KFC and PETA after disinterested investigation: PETA is a radical group, maybe even a dangerous one, and its claims should be regarded with skepticism. But just because it says something, doesn't mean it's not true. KFC, like most companies, blows a little corporate smoke. Its interest is the bottom line, not the well-being of chickens. But it is far from a nefarious company; it's just another chicken buyer. PETA may force the more humane treatment of chickens, which would be splendid. But the business of serving chicken, and other meats, to many millions of customers will always be a little dirty, something from which sensitive people rather turn away, even as they tolerate it, and benefit from it.

The PETA president, Ingrid Newkirk, has talked of finding Colonel Sanders's grave and dancing on it. Her group has promised that the company is "in for a long battle." If the company has a case to make—and it does—it ought to fight back, and hard. For its foe is formidable, and well-wishers of both KFC and animals could use a little honest reassurance.

RESTAURANTS SHOULD FIGHT PETA'S ANTI-MEAT AGENDA

Megan Steintrager

Megan Steintrager writes for *Restaurant Business* magazine. In the following selection, she reports on the influence of animal rights groups such as People for the Ethical Treatment of Animals (PETA) on the restaurant business. Steintrager argues that activist groups have caused unprecedented changes in the restaurant industry. Groups like PETA are forcing large chain restaurants to make changes in animal welfare guidelines, partly in order to avoid negative publicity, Steintrager writes. However, she reports, the restaurants claim that they are not caving in to the activists—instead, they insist they are choosing to set their own standards for animal welfare. Nevertheless, Steintrager asserts, restaurants are giving in to the activists, whose ultimate goal is to eliminate meat from restaurants entirely. She concludes that restaurant operators need to stick together in order to prevail over the activists.

The buzz was everything a restaurateur could ask for when the high-end Chicago sushi restaurant Heat opened [in] January [2002]. The restaurant garnered a lot of favorable reviews, including three out of four stars ("excellent") in the *Chicago Tribune*. "Fresh" was the word on everyone's lips and soon the restaurant was packed with people clambering to taste the creations of chef-owner Kee Chan and his staff.

Then the letters and phone calls started. Unfortunately for Chan, those weren't from fans.

News of Heat's fresh sashimi had reached animal rights activists, and several of the restaurant's dishes were a bit too fresh for them. In fact, some of it was still kicking—or to be more accurate, twitching and flopping—when it reached the table. Patrons were encouraged to behead live sweet prawns before peeling and eating the tail, for example. Fish arrived at the table fully alive—with chunks cut out of the belly.

People for the Ethical Treatment of Animals (PETA), probably the best known and certainly the most influential animal-rights group in the U.S., alerted its members of these "atrocities" at Heat and Mirai, another Chicago sushi restaurant, and hundreds of letters were writ-

Megan Steintrager, "Duty and the Beast," *Restaurant Business*, vol. 101, June 15, 2002, p. 20. Copyright © 2002 by VNU Business Media. Reproduced by permission.

ten to the Anti-Cruelty Society in Chicago and the Illinois Bureau of Welfare. A commentary published on PETA's web site likened dinner at these restaurants to how we'd feel if "Hannibal Lecter decided to hack off one of our legs for a midnight snack."

Chan says he tried to explain to the activists who approached him that it's a common practice in Japan to serve live fish, or fish that's killed tableside, to demonstrate the quality of the product. He also pointed out that raw oysters and clams are also technically alive when they're eaten. Attempting to bring the argument home for consumers, Chan asked rhetorically: "If you're going fishing, what is the first thing you are going to do with the fish? Smash his head and put it on the grill."

Chan's arguments fell on deaf—or disgusted—ears. The Bureau of Animal Welfare ordered Heat and Mirai to stop what they were doing, and both complied. PETA announced victory in the "Success Stories" section of its web site. Saying that chopping the heads off aquatic creatures behind the scenes hasn't hurt his business—or the freshness of the fish, Chan concludes, "We live in the United States and we have to make everybody happy."

The Influence of Activists

What does one trendy restaurant slightly altering its offerings have to do with the rest of the restaurant industry? As it turns out, a lot. As insignificant as this animal rights victory might seem to everyone but Chan and a small group of live-sashimi lovers, it's a telling example of the unprecedented headway that animal welfare activists have made throughout the industry in the past few years.

Though top brass are loath to admit a direct connection, McDonald's, Burger King, and Wendy's all made changes to their animal welfare policies following campaigns by PETA—a group that openly and loudly states that "animals are not ours to eat, wear, experiment on, or use for entertainment." Its mission isn't just vegetarian, it's vegan—condemning the consumption of any animal products, including eggs, milk, and cheese.

What's more, Burger King and the National Council of Chain Restaurants (NCCR) are now promoting even stricter guidelines and monitoring of the way food animals are raised and slaughtered. Gone are the days when operators could depend on the public to dismiss animal-rights groups as fringe oddballs, or simply claim animal welfare is a supplier issue.

Banding Together

For many restaurateurs, particularly the large chains who are easy targets for protestors, animal welfare has climbed up to the top of the public-policy priority list, alongside crime, accusations of promoting obesity, debates over smoking in restaurants, and threats of bioterrorism. With simply ignoring the situation no longer an option, opera-tors tried just about everything, including—as with Burger King and NCCR—working for some of the same results as the animal rights activists themselves. As with those other hot-button issues, a growing number of operators have decided that the best way to deal with the activists is to band together as an industry.

Animal-rights groups are finally forcing the hands of restaurants and restaurants are quietly going along. Large chains are demanding audits of slaughterhouses to ensure that existing laws (like the one that states that cows and pigs must not be dismembered while still alive) are being obeyed and they're calling for things like more room for chickens to move, and the end to the practice of forcing egg pro-duction by cutting off hens' food.

Because such moves have clearly been low-key, you might say that restaurant chains appear to be playing a quiet game of If You Can't Beat 'Em, Join 'Em. On the surface, that might not seem like a bad approach for an operator. After all, recent stances taken by the indus-try have good PR value. They also take the pressure off restaurants, shifting reform responsibilities onto the shoulders of suppliers. Then again, it could also be making a pact with the devil. Animal-rights groups like PETA have a clear vegan agenda they've vowed to never abandon, and the industry could never acquiesce to that. Restaurants may be willing to satisfy today's demands, but how about tomor-row's? The situation poses an unpleasant question: Could this whole thing blow up in the industry's face?

Playing into PETA's Hands

In recent years, several major chains have played right into the hands of activists, even those activists who openly admit their divide and conquer approach.

"We can smear their symbols so that people choose to eat else-where," explains Bruce Friedrich, PETA's Vegan Campaign Coordina-tor. "We would not be nearly as successful in the mainstream if we told people 'don't eat anywhere.'"

To that end, in 1999, PETA called off two years of "negotiations" with McDonald's and launched its McCruelty Campaign, which included noisy demonstrations at restaurants; a McCruelty web site featuring a bloodthirsty "Son of Ron" mascot; and billboards that fea-tured a bloody, dismembered cow's head along with the slogan, "Do you want fries with that?"

When McDonald's came out with its new standards, PETA moved the show to Burger King ("Murder King,"), then to ("Wicked") Wendy's, where the group kicked off their protests at a store in Vir-ginia at which Babe star James Cromwell was arrested. Even PETA's more humorous tactics—members dressing up like fuzzy farm animals or stripping down to their birthday suits—seem to be effective. Like

McDonald's, both Burger King and Wendy's announced new animal-welfare guidelines immediately following PETA campaigns. Now PETA has announced that they've begun writing letters to pizza companies demanding they force suppliers to upgrade their standards or risk similar treatment.

Caving to Activists

Some see such acquiescence on the part of chains to be little better than a sign of weakness. "The industry is so competitive that chains play into the tactics of PETA, one at a time," grumbles Rick Berman, who heads up the Center for Consumer Freedom (formerly the Guest Choice Network), a group that forcefully advocates the public's right to smoke, drink, and eat as it pleases. "There are people who feel that some of the companies should have taken a stronger stand," he says.

Janet Riley, VP of public affairs for the American Meat Institute, says PETA has "come up with a fairly effective formula" that has "raised the profile of the issue." She worries that the group provoked some companies to "out humane" each other. "It can start to raise the stakes," she adds. "It's a slippery slope."

The stakes were raised [in] June [2001] when Burger King announced what they called "industry-leading guidelines," which will require the company's suppliers to "adhere to the strictest standards" of animal welfare.

Did Burger King cave to activists? According to senior communications VP Rob Doughty, these measures were simply a result of the research and suggestions from its own Animal Well-Being Advisory Council, which is comprised of scientists, animal handling experts, and BK staffers.

"We had been watching the issue for some time," he says, denying that the protestors spurred them on. "I know that they [PETA] make it look like we had serious conversations with them," he adds. "We have chosen to be a leader in setting our own standards." Regardless of the impetus, Doughty says the company "changed our thinking."

Strengthening Animal Welfare

Indeed, the industry has come around to animal-rights groups' ways of thinking through more than just the policy actions of one major chain. Recently, the National Council of Chain Restaurants and the Food Marketing Institute assembled an advisory panel to "identify issues and weaknesses" in current animal welfare standards. In late February 2002 the group released an interim report "on efforts to further develop and support food industry programs that strengthen animal welfare." The group's formation was a move that was "certainly driven by some publicity put on this by some activists groups," says the Council's president Terrie Dort.

But she insists that the steps were primarily motivated by revelations that "there are some things going on that shouldn't be going on." In addition, she says that informal surveys indicated that consumers—even meat eaters—cared about the issue, and would push for stricter animal welfare if they learned about mistreatment of farm animals.

In discussing treatment, companies are favoring what's known as "science-based" standards—something they're much more comfortable with than PETA's stomach-turning descriptions of pregnant sows stuffed into cages and chickens with their beaks chopped off. As for PETA, Dort says it would be counterproductive to put people with a vegan agenda on the panel, but adds, "We see all of their information—we are aware of what they want and what their concerns are."

Guided by other catchphrases like practical, tangible, measurable, and cost-effective, the panel is now also working with producers to come up with standards that are mutually agreeable for producers and operators—what McDonald's senior director of social responsibility Bob Langert calls "a credible approach." Burger King, McDonald's, and the NCCR say that producers have been eager to work with them, as they were already working on new guidelines and procedures themselves. But eager though they may be, suppliers have little choice in the matter. By siding with the activists, the chains are forcing suppliers hands—they either go along or risk looking like animal torturers.

Producers are anxious to have a hand in shaping the guidelines that would be imposed on them by their restaurant customers—especially since compliance isn't exactly optional. "If they don't do it voluntarily, it will be mandated on them by their biggest customers," says Dort.

Restaurants Should Stick Together

But the question that keeps rearing its head for suppliers and operators alike is where does it end? Dort hopes that an industry audit will be enough to reassure mainstream customers that adequate standards are being met and that they can dig into their meaty meals with clear consciences. After all, meat is still a very popular menu item in the U.S. But placating mainstream customers is one thing; satisfying PETA is something else. As Doughty puts it: "We have no intention of becoming Veggie King."

Yet veritable Veggie Kings are exactly what groups like PETA want. While endorsing the changes that have been made thus far—including BK's new Veggie burger—PETA's Friedrich says the organization is still advocating further concessions from restaurant companies, including those that have already made changes (a PETA web page depicts Ronald McDonald behind bars and announces that McDonald's is "on probation"). He won't be satisfied until "no corporations are serving up animal products."

How serious a threat is that? It depends who you ask. Berman's Center for Consumer Freedom thought it was serious enough to

mount an ad campaign that features some of PETA's more inflamma-
tory rhetoric in an attempt to portray them as extremists to the public. One ad quotes Friedrich making the statement: "It would be great
if all the fast food outlets, slaughterhouses, these laboratories, and the
banks that fund them exploded tomorrow." The group's web site dis-
courages consumers from supporting PETA by drawing links between
the group and violent animal rights and environmentalist groups—of
course, PETA denies the connections.

Left- and right-wing histrionics aside the worst scenario for operators
and producers is the best scenario for for the activists—a failure on the
part of the industry to continue to act together. McDonald's Langert
stresses that the more operators and producers that agree to adopt certain standards, the more efficient the entire process will be. Seconds
BK's Doughty: "The food industry would do best to develop a common
set of standards so that everybody is on the same playing field." In
other words: don't divide and you're less likely to be conquered.

Consumers Are Encouraging Humane Food Production

Laurent Belsie

In the following selection, journalist Laurent Belsie explains that consumer groups are responsible for a trend away from factory farming of food animals such as chickens and pigs. Consumers are interested in the welfare of animals, Belsie notes, and food producers are responding by implementing new standards of safe and humane production. The response by restaurant giant McDonald's—which has improved the conditions under which its supplier's chickens live—has led others in the industry to begin following suit, Belsie reports. However, the author relates, the humane treatment of animals requires tradeoffs, such as increased production costs that are passed on to consumers. Belsie is a department editor for the *Christian Science Monitor.*

Farmers are productivity stars. They keep finding ways to produce cheaper and fresher food. But in their push to raise animals more efficiently, American farmers find themselves in a thicket of ethical problems.

Their production methods could be hurting the animals they raise—and, in some cases, clearly are.

While several European nations have restricted some of the most controversial practices, here in the United States the private sector is leading the efforts at reform. If a handful of food companies and non-govemmental organizations are successful, their initiatives could force American agriculture to take its closest look yet at animal welfare and the drawbacks of conventional farming.

A small contingent of consumers already are paying close attention. And while they currently don't have the market clout to transform the livestock industry, their numbers are growing.

"There's genuinely a new ethic emerging for animals in society," says Bernard Rollin, director of bioethical planning at Colorado State University in Fort Collins, Colo.

"This is the issue that consumers are going to be concerned about,"

Laurent Belsie, "Chicken Tenderly," *Christian Science Monitor*, vol. 92, November 6, 2000, p. 11. Copyright © 2000 by The Christian Science Publishing Society. All rights reserved. Reproduced by permission.

adds Adele Douglass, executive director of Farm Animal Services, a nonprofit certification organization created by the American Humane Association (AHA) in Washington. The movement has begun.

Certifying Humane Treatment

In September [2000], for example, the AHA introduced a "Free Farmed" label for food products that meet its criteria on humane treatment. Already, three small companies have won certification and many more are expected. "I have been inundated with calls from producers," says Ms. Douglass, who's fielding 25 e-mails and 20 phone calls a day.

The label should help consumers sort out how their meat, eggs, and milk was produced, which isn't always readily apparent. For example, meat labeled "natural" by the US Department of Agriculture (USDA) means the animal did not receive growth-enhancing antibiotics or hormones, but says nothing about its living conditions.

An "organic" label usually implies better conditions, but standards vary among the nearly 50 state and private organic-food certifiers. And federal labels for organic food, which include general language about animal welfare, [didn't] appear on shelves until 2002.

The fact that a consumer buys from a company that sells organic food doesn't mean he or she is backing a "humane practices" farm. Producers of naturally raised livestock sometimes run conventional operations, too—often out of necessity.

In their "natural" operations, they don't feed animals small amounts of antibiotics to make them grow faster. But they will use antibiotics to treat disease. And if a treated animal gets better, many of these producers will raise it conventionally rather than kill it or try to sell it as "natural."

The best course until the humane-labeling system is fully functioning? "Go to a grocer you trust," says Paul Gingerich, meat and seafood director for Wild Oats Markets, a fast-growing organic grocery chain with stores in the US and Canada. And if the grocer doesn't know, contact the manufacturer, he adds.

McDonald's Makes Good

One company that has already responded to demands by animal activists is McDonald's. The world's largest fast-food chain, based in Oak Brook, Ill., established in August [2002] humane guidelines for egg production. Its biggest change: asking its suppliers to increase by one-third the cage space they allot for each laying hen.

"We're improving the lives of 5 million hens," says Bob Langert, McDonald's senior director of public affairs who heads up environment and animal welfare. "We're committed to animal-welfare leadership in the long term. I think there's more to come."

Although the McDonald's guidelines are not as far-reaching as those of Farm Animal Services—which include requirements for perches and sandlots for the hens—the company's sheer size is causing other large food processors and restaurant chains to take notice. At least one animal-rights group, People for the Ethical Treatment of Animals, is pressuring Burger King to follow McDonald's lead. In a statement, the fast-food chain said it was evaluating the egg standards used by McDonald's.

"I think we're going to see a lot more private-sector, retail-driven" initiatives, says Jeff Armstrong, head of the animal-sciences department at Purdue University in West Lafayette, Ind. "If that's successful, we will not need government regulations."

Some poultry experts say the changes are long overdue.

For example, the egg industry currently allots the average hen a 6-inch by 9-inch square of cage space. All the hens can't sit down at the same time. McDonald's guidelines, which closely mirror the voluntary code recently set by the United Egg Producers, a large egg-industry group, eliminate that problem by increasing the allotted cage space to 72 square inches per bird.

Other experts want the industry to go much further. "There is a better way to treat chickens," says John Brunn-quell, founder of Egg Innovations. His Port Washington, Wis., company raises free-roaming hens without resorting to artificial hormones. And to become the first egg company certified with the Free Farmed label, the company spent tens of thousands of dollars to install sand lots and perches so hens could engage in other natural activities, such as dust-bathing.

Consider the Tradeoffs

But there are tradeoffs. For one thing, open systems don't isolate hens from bacteria as well as cages do. And either way, beaks need to be trimmed because otherwise, big flocks of hens are known to engage in cannibalism.

That's why noted poultry experts, such as Joy Mench, director of the Center for Animal Welfare at the University of California at Davis, are wary of the cageless systems common in Europe. "The more extensive production systems for laying hens really haven't worked very well," she says.

Another huge bugaboo is cost. Although McDonald's says it's premature to say how much its guidelines will raise prices, industry estimates range from 15 to 40 cents a dozen. Brand-name niche companies, such as Egg Innovations, can meet the cost by charging double for a product. But it's unclear whether consumers of commodity eggs will accept a 50 percent price hike.

"Producers genuinely want to do the right thing; however, they also have to turn a profit," says Professor Armstrong of Purdue. He thinks a market-driven approach can achieve both goals. A 1999 survey by the Animal Industry Foundation found that 44 percent of Americans would pay 5 percent more for meat and poultry that was "humanely raised."

So far, most consumers don't shop that way. Organic food remains a tiny part of the food business—about 1 percent in the US, Europe, and Japan, according to the United Nations' International Trade Center. But its share could grow to 10 percent in the next few years, the center says in a new report. In the US, organic food has become the fastest growing retail business, says Mr. Gingerich of Wild Oats.

Eggs aren't agriculture's only ethically tricky product. Raising chickens for meat also poses problems. To begin with, broiler chickens (as they're called) are bred to grow so quickly that near the end of their short lives their bodies can't properly support the weight they gain. The result? "Ninety percent of them walk abnormally," Dr. Mench says. "About 20 percent have really severe mobility problems."

Some small producers have dumped the factory model and market free-range poultry instead. In good weather, the chickens go outside. But those producers have to charge considerably higher prices than large commercial operators. So instead of closing their facilities, commercial operators are looking at various ways to improve breeding, alter growth cycles through lighting, and modify cages so hens can perch and dust-bathe. So far, though, the broiler industry has not developed guidelines.

Porcine Predicament

Another ethical dilemma involves hog production. Currently, most of the pork Americans eat comes from animals who spend their lives in small, indoor pens. The smallest of these units, measuring two feet by seven feet, contain pregnant sows. For more than 100 days, the sow is isolated, can't turn around, and can't play.

Confinement confers advantages. Farmers can treat and track individual animals more easily. Timid sows are protected from more aggressive hogs who occasionally bite. And hogs generally grow faster and more consistently in a controlled indoor environment.

But is confinement ethical?

Out in the wild, a sow might forage for food up to a mile away, points out Professor Rollin, the bioethicist. "To put an animal like that and put it in a box smaller than it is, to me is purely evil. It's a matter of going too far for the sake of pecuniary considerations."

To answer the criticism, the National Pork Producers Council convened an international panel of experts to review more than 800 research articles on the subject. Their conclusion: There are no clear behavioral or bodily clues that the sow is under stress in confinement.

"We can't show that she's thinking anything other than 'I'm going to eat and drink,'" says Paul Sundberg, assistant vice president for veterinary issues for the council, based in Des Moines, Iowa. "The person taking care of the animals is probably more important than the system of housing."

The council's animal-welfare committee will reevaluate its producers code and may address some problems, such as making sure that the confinement units are adjusted to the size of the sow. But "my sense is that the science doesn't give us the basis to make wholesale changes at this time," Dr. Sundberg says.

Less to Beef About

Of all production farm animals, dairy and beef cattle retain the most freedom. But even here, concerns remain about conditions in feedlots, veal farms, and dairy operations where cows don't have enough stalls, and stand on concrete for long hours. Some small producers are marketing alternatives.

"We know what we're doing is right for this time," says Dan Benedetti of Clover-Stornetta Farms. The private dairy marketer in Petaluma, Calif., sells milk from farmers who certify that their product comes from cattle that graze in open pastures and aren't injected with a controversial growth hormone.

"I'm not sure it's a niche anymore," he adds. "That whole market segment is growing."

EATING MEAT IS MORALLY INDEFENSIBLE

David DeGrazia

David DeGrazia is associate professor of philosophy at George Washington University. In this excerpt from his book *Animal Rights: A Very Short Introduction*, DeGrazia makes the case that animals raised for food undergo a tremendous amount of unnecessary suffering. For this reason, he maintains, it is immoral to eat meat. Consumers do not need the products produced by factory farms, DeGrazia argues, which means that there is no way to justify the harm that farm animals suffer. In fact, he asserts, eating meat is bad for human health. Not only that, DeGrazia continues, factory farming degrades the environment and contributes to the unequal distribution of global resources, resulting in more human suffering. DeGrazia holds that ending meat consumption is the only way to stop unnecessary harm to farm animals.

Hen X begins life in a crowded incubator. She is taken to a 'battery' cage made entirely of wire—and quite unlike the outdoor conditions that are natural for her—where she will live her life. (Having no commercial value, male chicks are gassed, ground up alive, or suffocated.) Hen X's cage is so crowded that she cannot fully stretch her wings. Although her beak is important for feeding, exploring, and preening, part of it has been cut off, through sensitive tissue, in order to limit the damage caused by pecking cage mates—a behaviour induced by overcrowding. For hours before laying an egg. Hen X paces anxiously among the crowd, instinctively seeking a nest that she will not find. At egg-laying time, she stands on a sloped, uncomfortable wire floor that precludes such instinctual behaviours as pecking for food, dust bathing, and scratching. Lack of exercise, unnatural conditions, and demands for extreme productivity—she will lay 250 eggs this year—cause bone weakness. (Unlike many hens, Hen X is not subjected to forced moulting, in which water is withheld one to three days and food for up to two weeks in order to extend hens' productive lives.) When considered spent at age 2, she is jammed into a crate and transported in a truck—without food, water, or protection from the elements—to a slaughterhouse; rough handling causes several weak bones to break. At her destination, Hen X is shackled upside down on a conveyor belt before an automated knife slices her throat. Because the (US) Humane Slaughter Act does not apply to poultry, she is fully conscious throughout this process. Her body, which was extensively damaged during her lifetime, is suitable only for pot pies, soup, and the like.

Factory Farms Cause Suffering

After weaning at four weeks of age, Hog Y is taken to a very crowded, stacked nursery cage. Due to poor ventilation, he breathes in powerful fumes from urine and faeces. Upon reaching a weight of 50 pounds, he is taken to a tiny 'finishing' pen. It is slatted and has a concrete floor with no straw bedding or sources of amusement. Despite being a member of a highly intelligent and social species, Hog Y is separated from other hogs by iron bars and has nothing to do except get up, lie down, eat, and sleep. He sometimes amuses himself by biting a tail in the next crate—until all the hogs' tails are 'docked' (cut off). Both this procedure and castration are performed without anaesthesia. When he is deemed ready for slaughter, Hog Y is roughly herded into a truck with thirty other hogs. The two-day journey is not pleasant for Hog Y, who gets in fights with other hogs while receiving no food, water, rest, or protection from the summer heat. At the slaughterhouse, Hog Y smells blood and resists prodding from the human handlers. They respond by kicking him and smashing him repeatedly from behind with an iron pipe until he is on the restraining conveyor belt that carries him to the stunner. Hog Y is fortunate in so far as the electric stunning procedure is successful, killing him before his body is dropped in scalding water and dismembered. (Although the Humane Slaughter Act requires that animals other than poultry be rendered unconscious with a single application of an effective stunning device before being shackled, hoisted upside down, and dismembered, many slaughterhouse employees state that violations occur regularly. Fearing that a higher voltage might cause 'bloodsplash' in some carcasses, many slaughterhouse supervisors apparently encourage use of a voltage that is much too low to ensure unconsciousness. Moreover, in numerous slaughterhouses stunners have to stun an animal every few seconds and face extreme pressure not to stop the line of animals.)

Although it is natural for cows and their calves to bond strongly, Cow (then Calf) Z is taken from her mother shortly after birth to begin life as a dairy cow. She never receives colostrum—her mother's milk—which would help her fight disease. She lives in a very crowded 'drylot', which is devoid of grass, and her tail is docked without anaesthesia. In order to produce twenty times more milk than a calf would need, she receives a diet heavy in grain—not the roughage that cows have evolved to digest easily—causing metabolic disorders and painful lameness. And like many dairy cows, she often has mastitis, a painful udder inflammation, despite receiving antibiotics between lactations. To maintain continuous milk production, Cow Z is induced to bear one calf each year. To stimulate additional growth and productivity, she receives daily injections of bovine growth hormone. Her natural life span is twenty or more years, but at age 4 she can no longer maintain production levels and is deemed 'spent'. During transport and handling, Cow Z is fortunate: although deprived of food, water, and rest for over two days, and frightened when prodded, she is not beaten; at the slaughterhouse her instincts—unlike hogs'—allow her to walk easily in a single-file chute. Unfortunately, the poorly trained stun operator has difficulty with the air-powered knocking gun. Although he stuns Cow Z four times, she stands up and bellows. The line does not stop, however, so she is hoisted up on the overhead rail and transported to the 'sticker', who cuts her throat to bleed her out. She remains conscious as she bleeds and experiences some of the dismemberment and skinning process alive. (The federal inspector cannot see what is happening where he is stationed; besides, he's frenetically checking carcasses that whiz by, for obvious signs of contamination.) Cow Z's body will be used for processed beef or hamburger.

The Institution of Factory Farming

The animals portrayed above offer examples of life in modern factory farms, which now supply most of our meat and dairy products in the USA, Great Britain, and most other industrial countries. Since the Second World War, factory farms—which try to raise as many animals as possible in very limited space in order to maximize profits—have driven three million American family farms out of business; over the same time period, Great Britain and other nations have witnessed similar transformations in their agricultural sectors. Scientific developments that have fuelled the emergence of factory farming include the artificial provision of vitamin D (which otherwise requires sunlight for its synthesis), the success of antibiotics in minimizing the spread of certain diseases, and advanced methods of genetic selection for production traits. Since the driving force behind this institution is economic efficiency, factory farming treats animals simply as means to this end—as mere objects with no independent moral importance, or moral status, whatever.

Considering both numbers of animals involved and the extent to which they are harmed, *factory farming causes more harm to animals than does any other human institution or practice.* In the USA alone, this institution kills over 100 million mammals and five billion birds annually. American farm animals have virtually no legal protections. The most important applicable federal legislation is the Humane Slaughter Act, which does not cover poultry—most of the animals consumed—and has no bearing on living conditions, transport, or handling. Moreover, as Gail Eisnitz and others have extensively documented, the Act is rarely enforced. Apparently, the US Department of Agriculture supports the major goal of agribusiness: absolute maximization of profit without hindrance. This is not surprising when one considers that, since the 1980s, most top officials at USDA either have been agribusiness leaders themselves or have had close political and financial ties to the industry.

By contrast, European nations have curbed some of the excesses typified by American factory farming. For example, Great Britain has banned veal crates and limits to fifteen hours the amount of time animals can go without food and water during transport. The European Community and the Council of Europe have developed requirements for the well-being of farm animals that are translated into law in different member nations. These requirements generally provide animals with more space, greater freedom to engage in species-typical behaviours, and more humane living conditions than those of farm animals in the USA. Despite the more humane conditions that are typical in Europe, however, most European animal husbandry remains sufficiently intensive to merit the term 'factory farming'.

All Farm Animals Suffer

So far this discussion has provided a descriptive sense of factory farms primarily through three cases. Therefore it might be objected that the situations of Hen X, Hog Y, and Cow Z do not represent universal features of factory farming. That is correct. But the experiences of these three animals, the evidence suggests, are not atypical—at least in the USA. Still, while a thorough description of factory farms is impossible here, it may be helpful to add a few general remarks about other types of farm animals. The following generalizations are meant to describe the American situation, although some of them accurately describe the experiences of animals in many other countries as well.

Cattle raised specifically for beef are generally better off than the other animals described here. Many have the opportunity to roam outdoors for about six months. After that, they are transported long distances to feedlots, where they are fed grain rather than grass. Major sources of pain or distress include constant exposure to the elements, branding, dehorning, unanaesthetized castration, the cutting of ears for identification purposes, and a sterile, unchanging environment. We may add, of course, the harms associated with transportation to the slaughterhouse and what takes place therein.

Broiler chickens spend their lives in enclosed sheds that become increasingly crowded as tens of thousands of birds grow at an abnormally fast rate. Besides extreme crowding, major sources of concern include cannibalism, suffocation due to panic-driven piling on top of one another, debeaking, and very unhealthful breathing conditions produced by never-cleaned droppings and poor ventilation. Veal calves' deprivations are similar to many of those that hogs experience. Formula-fed veal calves in particular live in solitary crates too small to permit them to turn around or sleep in a natural position. Denied water and solid food, they drink a liquid milk replacer deficient in iron—making possible the gourmet white flesh and resulting in anaemia. This diet and solitary confinement lead to numerous health problems and neurotic behaviours.

Let us now consider the overall picture: *factory farming routinely causes animals massive harm in the form of suffering, confinement, and death.* Regarding suffering—or experiential harm in general—all evidence suggests that factory farm animals, in the course of their lives, typically experience considerable pain, discomfort, boredom, fear, anxiety, and possibly other unpleasant feelings. Furthermore, factory farms by their very nature *confine* animals in our stipulated sense of the term; that is, they impose external constraints on movement that significantly interfere with living well. (For at least part of their lives, cattle raised specifically for beef are not confined in this sense.) And, of course, factory farming ultimately kills animals raised meat, adding the harm of death—assuming . . . that death harms such beings as cows, pigs, and chickens. Then again, death counts as a harm here only if we consider the sorts of lives these animals *could* have under humane treatment. Given animals' current treatment, death would seem to be a blessing, except possibly in the case of beef cattle. In any event, the general thesis that factory farms cause massive harm to animals is undeniable.

Moral Evaluation

If the first crucial insight in a moral evaluation of factory farms is that they cause massive harm to animals, the second crucial insight is this: *consumers do not need the products of factory farms.* We cannot plausibly regard any of the harms caused to these animals as *necessary.* Unusual circumstances aside—say, where one is starving and lacks alternatives—we do not need to eat meat to survive or even to be healthy. The chief benefits of meat-eating to consumers are *pleasure,* since meat tastes especially good to many people, and *convenience,* since switching to and maintaining a vegetarian diet requires some effort. Putting the two key insights together brings us to the conclusion that *factory farms cause massive unnecessary harm.* Since causing massive unnecessary harm is wrong if *anything* is wrong, the judgement that factory farming is an indefensible institution seems inescapable.

Note that this condemnation of factory farming does not depend on the controversial assumption that animals deserve equal consideration. Even if one accepts a sliding-scale model of moral status, which justifies less-than-equal consideration for animals, one cannot plausibly defend the causing of massive unnecessary harm. Thus, it appears that if one takes animals at all seriously—regarding them as beings with at least some moral status—one must find factory farming indefensible.

Meat Consumers Support Animal Suffering

But what about the consumer? She isn't harming animals; she's just eating the products of factory farming. Well, imagine someone who says, 'I'm not kicking dogs to death. I'm just paying someone else to do it.' We would judge this person to act wrongly for encouraging and commissioning acts of cruelty. Similarly, while meat-eaters may typically feel distant from meat production, and may never even think about what goes on in factory farms and slaughterhouses, the purchase of factory-farmed meat directly encourages and makes possible the associated cruelties—so the consumer is significantly responsible. In general, the following moral rule, although somewhat vague, is defensible: *make every reasonable effort not to provide financial support to institutions that cause extensive unnecessary harm.*

By financially supporting massive unnecessary harm, the purchase of factory-farmed meat violates this principle and is therefore, I argue, morally indefensible. Interestingly, we reached this important conclusion without commitment to any specific ethical theory such as utilitarianism or a strong animal-rights view. In any event, while our case against factory farming and buying its products has so far cited considerations of animal welfare, it is further strengthened by considerations of human welfare. How so?

Meat Consumption Is Bad for People

First, animal products—which are high in fat and protein and contain cholesterol—are associated with higher levels of heart disease, obesity, stroke, osteoporosis, diabetes, and certain cancers. Medical authorities now recommend much less meat and more grains, fruits, and vegetables than Americans, for example, typically consume. Second, American factory farming has driven three million family farms out of business since the Second World War, as huge agribusinesses, enjoying billions of dollars in annual government subsidies, have increasingly dominated; while American consumers frequently hear that factory farming lowers meat prices at the cash register, they are rarely reminded of the hidden cost of tax subsidies. In Britain and many other countries, relatively few large agribusinesses have similarly come to dominate, putting many smaller farms out of business. Third, factory farming is devastating for the environment. It excessively consumes energy, soil, and water while causing erosion of topsoil, destruction of wildlife habitat, deforestation, and water pollution from manure, pesticides, and other chemicals. Fourth, factory farming has a perverse effect on the distribution of food to humans. For example, it takes about 8 pounds of protein in hog feed to generate one pound of pork for humans and 21 pounds of protein in calf feed to yield 1 pound of beef. Consequently, most US-produced grain, for example, goes to livestock. Unfortunately, wealthy countries' demand for meat makes plant proteins too costly for the masses in the poorest countries. Poor communities often abandon sustainable farming practices to export cash crops and meat, but profits are short-lived as marginal lands erode, causing poverty and malnutrition. There is, in fact, easily enough grain protein, if used sensibly, to feed every human on Earth. Fifth, perhaps especially in the USA, factory farming is cruel to its employees. It subjects them to extreme work pressures—as seen in a worker who cuts up to ninety chickens per minute, or urinates on the workline for fear of leaving it—and to some of the worst health hazards faced by any American workers (for example, skin diseases, respiratory problems, crippling hand and arm injuries, injury from wild, improperly stunned animals)—all for low pay. Finally, deregulation of the American meat industry since the 1980s, combined with extremely fast production lines, have made it virtually impossible to ensure safe meat. As noted by Henry Spira, it has been estimated that contaminated chicken, for example, kills 2,000 Americans a year.

Boycotting Factory Farms

Thus, receiving further support from considerations of human welfare, the case for boycotting factory farm products is extremely powerful. But let us not ignore the following important objection. One might argue that the continuation of factory farming is economically necessary. Putting this industry out of business—say, through a successful boycott—would obviously be devastating for agribusiness owners, but would also eliminate many jobs and possibly harm local economies. These consequences, the argument continues, are unacceptable. Thus, just as factory farming is necessary, so is the extensive harm it inevitably causes to animals—contrary to my charge of massive *unnecessary* harm.

In reply, we may accept the factual assumption about likely consequences while rejecting the claim that they are unacceptable. First, as Peter Singer notes, the negative costs of ending factory farming would have to be borne only once, whereas perpetuating this institution entails that the costs to animals continue indefinitely. Also, considering how badly factory farm employees are treated, it is hard to believe they would be seriously harmed by having to seek alternative employment, as innumerable 'burn out' employees do anyway. More generally, the values threats to human well-being posed by factory farming—health risks, environmental destruction, inefficient use and perverse distribution of grain proteins, etc.—could be avoided if this industry is eliminated (assuming it is not simply replaced by less intensive animal husbandry, which would perpetuate some of these problems). Avoiding these risks and harms, not once but indefinitely, would seem to counterbalance any short-term economic harm. Finally,

I submit that *there are moral limits to what we may do to others in the pursuit of profit or employment—and causing sentient beings massive harm in pursuing these goals oversteps those bounds.* (Cases in which people are forced into prostitution, pornography, or slavery vividly exemplify the violation of such limits.) If that is correct, then factory farming cannot be considered necessary. In conclusion, I suggest that these rebuttals, taken together, undercut the argument from economic necessity.

Traditional Family Farming

[I have] focused on factory farms because most of the animal products we consume come from this source. But people also eat animals from other sources, including traditional family farms.

Because they involve far less intensive rearing conditions, family farms cause much less suffering to animals than factory farms do. Family farms may not even confine animals in our sense of imposing constraints on movement that significantly interfere with living well. But, assuming the opportunities-based account of the harm of death is correct, farm animals cannot fully escape harm because they are ultimately killed, entailing the harm of death.

Causing much less harm to animals, and avoiding at least some threats that factory farming poses to human well-being (for example, water pollution, extremely hazardous working conditions), family farming is much more defensible than its dominant competitor. Still, there is a strong moral case against family farming and the practice of buying its products. For one thing, this institution does impose some significant suffering through certain practices: branding and dehorning cattle; castrating cattle and hogs; separating mothers from offspring, which may well cause distress even to birds; and treating animals roughly in transport, handling, and slaughter. And, again, all the animals die. Since meat-eating is—unusual circumstances aside—unnecessary, these harms are unnecessary. It is difficult to defend the routine imposition of unnecessary harm.

A few possible replies, however, may strengthen the case for some forms of family farming. For example, chickens and turkeys can escape most of the harms just described. If a chicken or turkey is able to live a pleasant life—say, with family intact—and is never abused, the only relevant harm would be death. But one who defends a desire-based account of the harm of death would probably deny that birds are even subject to this harm, suggesting that in optimal conditions poultry are not harmed at all.

Alternatively, if one (unlike the present author) accepts the sliding-scale model of moral status, one would grant unequal moral weight to the interests—including the avoidance of suffering—of different beings depending on their cognitive, emotional, and social complexity. Perhaps proponents of this ethical framework would defend practices of family farming that keep the admittedly unnecessary suffering to a minimum. They might argue that it is not always wrong to cause *minimal* unnecessary harm, even to mammals, especially if there are some significant benefits such as employment for farmers. Then again, one would need to consider negative effects on human welfare, such as extremely inefficient use of grain protein, in assessing the plausibility of this line of argument.

HOW EDUCATION ABOUT ANIMAL WELFARE IS CHANGING ATTITUDES ABOUT EATING MEAT

Scott McKeen

Does a growing concern for animal welfare mean that people will have to stop eating meat? *Edmonton Journal* staff writer Scott McKeen explores this question in the following selection. Many people support laws that protect animals from cruelty, McKeen notes, yet they continue to eat meat and use animal products. Animal rights activists have had a definite impact on the public, the author observes, as more people try vegetarianism and hunting rates decline. Yet at the same time, McKeen points out, most consumers make little connection between the meat on their plate and how it got there. Experts suggest that people should learn more about where their meat comes from and how it is produced, so they can make informed decisions about what they should eat.

Paradigm shifts and cultural revolutions are out of keeping with the bygone charm of Max's Light Cuisine, a hole-in-the-wall diner in south Edmonton, [Canada].

Max Pieroelie works the kitchen alone, fashioning steaming, delicious plates for the elbow-to-elbow dining room, which can hold 22 diners, but not 23.

But look past the chattering patrons, the mugs of coffee and wedges of apple pie—look just above the three-stool counter—and the unique character of the place reveals itself, along with Max's small role in an emerging revolution.

The chalkboard specials today include tofu pie, veggie burger and an enchilada stuffed with tempeh, or fermented soy beans. Lentil nut loaf, vegetable-tofu stir fry and black-bean chili rollup are everyday offerings at Max's, one of a growing number of restaurants in Edmonton catering to vegetarians.

Health issues, mostly, are driving this small but noticeable shift away from meat eating in our culture. But so is a growing concern about the welfare of the planet and the animals who share it with us.

Scott McKeen, "Moving Away from Meat," *Edmonton Journal*, June 16, 2002, p. D6.

Changing Attitudes About Animals

Animal rights were given a mammoth push [in June 2002, in Canada] when the House of Commons approved a new animal cruelty bill. While most Canadians likely missed the moment, entire industries gulped in unison at the news.

In the past, Canadian law has always treated animals as chattel—as inanimate objects without rights. But with the new legislation, animals have legal status as sentient beings with their own interests and values, worthy of protection as individuals, irrespective of ownership.

"It's a sea change in the way we look at animals," says Dave Neil, a University of Alberta professor of lab-animal science.

The legislation's aim is to give the courts more ammunition in dealing with animal abuse cases. [In spring of 2002], the two Toronto men who skinned a cat alive—ostensibly for "art"—were given relative wrist slaps that infuriated animal rights advocates. The new legislation provides for penalties of up to five years in jail.

While livestock producers, the fur industry and lab-animal scientists are worried about the new law, they have been assured the legislation . . . includes reasonable protection against frivolous prosecution.

It's conceivable an animal-rights group might use the . . . law to take a run at some scientist or rancher, but even with that possibility people like Dave Neil say the more compelling question raised by the bill is its implication for society as a whole.

With animals given new status as beings, instead of property, will public attitudes shift in that direction? Will it remain OK to use animals, when the law itself sees them as having intrinsic value?

Animal Rights and Eating Meat

Ranchers, lab scientists and trappers have been forced over the years to make their practices more humane. Perhaps the final frontier in the battle for animal rights is the kitchen table.

Mahatma Gandhi said a society's moral progress can be judged on its treatment of animals. Is it immoral, then, to even eat meat? Are we heading for a future of motherhood and tofu pie?

University of Alberta ecologist Lee Foote argues that morality is irrelevant to nature; that humans, as predators and meat eaters, are part of the natural order. Foote worries that vegetarianism and animal rights have gained an almost religious following—one which denies the right to personal choice and ignores the realities of nature.

"We are at a juncture in our cultural evolution, where we get a choice," says University of Alberta philosopher Margaret van de Pitte. "We can continue to treat animals as resources, or we might decide that we'll be happier if we no longer do."

Philosophy has grappled with such questons since its beginnings. Anthropocentrism—the view that humans are at the apex of existence—has dominated since the time of Aristotle, who believed animals were created by nature for man's benefit.

Christian belief largely followed that sentiment, as God was said to have given humankind dominion over the world and the beasts.

Rene Descartes considered the matter in the 17th century and argued that animals were like machines, incapable of thought or feeling. Cartesian thought was used widely to justify such things as experimentation on animals, without anesthesia. But it also infuriated another great French philosopher, Voltaire, who saw in animals a kinship with humans.

He challenged Descartes: "Answer me, machinist, has nature arranged all the means of feeling in this animal, so that it may not feel?"

English philosopher Jeremy Bentham went even further than Voltaire, arguing that animals were objects of legitimate moral consideration. Even though animals have no language, argued Bentham, they suffer. This capacity for suffering, he said, placed on humanity an obligation to be their moral agents.

Science in the 20th century revealed that animals do in fact communicate and that some species have language and use tools—things that were supposed to separate humans from lowly beasts.

The intelligence of dolphins and primates is well established, but even sheep and cows are now known to have complex social lives, suggesting cognitive and emotional capacity beyond what Bentham used for his moral argument.

While Britain established animal cruelty laws at about the time of Bentham's musings, Canada is just now catching up with the passing of the animal cruelty bill.

Social Concern for Animals

Van de Pitte believes, though, that Canadians remain deeply divided on the issue of animal rights. New law or not, countless legions of animals continue to be used in lab experiments, herded into the abattoir or skinned for their soft and pleasing fur.

Canadians also continue to eat meat in quantities unheard of in other places around the globe. Meat is plentiful, relatively inexpensive and marketed brilliantly in everything from the succulent filet mignon at the high-end steakhouse to the greasy burger at the fast-food joint.

But at the same time there is considerable evidence of a change in social consciousness, symbolized by the move by a multinational fast-food chain, Burger King, to introduce a veggie burger to its menu. Burger King is also at the forefront of a campaign to end the practice of branding cattle and has warned suppliers it will not buy their beef if cows are put to the brand.

This kind of social concern is driven by consumer demand, which is no doubt influenced by our high standard of living today. Affluence gives people not only the wherewithal to buy juicy steaks, but also the luxury of time to develop a social conscience.

Social concern is also heightened by the public's level of education these days, says David Neil, as well as the influence of mass media, which bombards us with information.

"You can't help but be informed these days," said Neil.

You also can't escape all the health news and how meat has been implicated in everything from diabetes and heart disease to allergies and cancer.

Even the seminal book *Baby and Child Care*, in its latest edition, recommends a vegetarian diet for children after age two. Its author, parenting guru Dr. Benjamin Spock, became a vegetarian later in life and believed kids raised without meat or dairy would enjoy tremendous health advantages.

Urban Attitudes About Animals

No doubt, urbanization in North America has also had a profound effect on the way the public views animal issues.

The fact that most of us live in cities, far from farms, results in a public almost completely disconnected from the reality of where their food comes from. For city dwellers, it's as if meat is a product grown on trees.

This urban-rural alienation plays out both in favour of animal rights and against them. City dwellers tend to personify animals—see them as smart and cute. But at the same time they don't make the mental connection between the meat on their plate and the death of the animal that provided it.

"The way we handle our food is designed to keep us from thinking about the real critters behind it," says van de Pitte.

Farmers often lament this knowledge gap, saying city folk are dismayed by practices like branding because they don't understand the reasons behind it, let alone the realities of raising livestock.

But the knowledge gap also plays in the cattle rancher's favour, according to van de Pitte. People who have rarely if ever visited a farm won't have a crisis of conscience—they won't have images of a cute lamb, calf or chick in their mind's eye—when they sit down to eat, she says.

Van de Pitte's own parents took her as a child to a slaughterhouse. She was deeply moved by the experience and today eats only vegetables and fish.

"I think anybody who goes to a slaughterhouse—who sees exactly what their food is like live, as opposed to a plasticized product—are put off their feed for a few days."

The Influence of Activists

If there has been a conscious effort by animal industries to hide the gory details from the public, there has been a reverse campaign by animal rights activists. This has had an impact on public attitudes, too.

The east coast seal hunt in Canada was put under siege, with copious gory videos. The same happened to the fur industry. A number of zoos around the world also closed their doors under this climate of growing social consciousness.

Perhaps the best known animal-rights group, People for the Ethical Treatment of Animals, or PETA, have staged numerous publicity campaigns, most of them outrageous. One involved naked celebrities proclaiming they'd rather be naked than in fur.

PETA even took on one of the most trusted foods of them all, milk, arguing that it wasn't the wholesome elixir it's been made out to be. The dairy industry responded with its own celebrity "wear a moustache" milk campaign.

Van de Pitte says the hunting industry also took a hit from changes in public mood, as well as considerable media attention. There has been a dramatic decline in recreational hunting, which she applauds, even though she thinks hunting is a "wildly more honest" way of acquiring meat than shopping at a grocery store.

But hunting, while a flashpoint with animal rights activists, has found supporters in the environmental movement. Hunters, say environmentalists, are wanderers of the wilderness and have become its strongest protector, as well as stewards of wildlife.

Eating Meat Is Part of the Natural Order

Ecologist Lee Foote is an ardent supporter of hunting and worries that Canada's . . . animal cruelty bill represents a further erosion of humankind's ties to the natural environment.

Foote sees hunting, and even meat eating, as a crucial part of the natural order—an order that sustains not only us, but the animals and the planet. He disagrees that people today would give up meat if they were forced to face the realities of killing and butchering animals.

"I would hope the exact opposite," said Foote. "I would hope that if they understood our origins, our adaptations, our predispositions, our evolution and our biology, they would see it as a beautiful, useful component of life."

Apologizing for citing Disney, Foote says there really is a circle of life. As predators, humans kill and eat animals, he says. But at some point, they die and their bodies decompose to return nutrients to the earth.

"We take and we give, each in its time," said Foote.

The circle of life and death is what people are really disconnected from, he says. Not just the fact that slaughterhouses are ugly, but that life is often ugly, just as it is beautiful. "There's a part of life that involves some suffering—suffering is part of death for people and animals alike," he says.

Foote doesn't believe in unnecessary suffering, but describes an almost pagan spirituality of respect for nature and its rules.

"Yes, it is spiritual," he says. "It's holistic and biological and practical, as well as spiritual.

"When somebody puts a bullet through a deer, an arrow through a moose or a bolt through a cow's head, these things don't play well to people who are disconnected from the whole process," he adds. "We as a society are scared to death of blood, and death. We seal our dead away under concrete and in fenced-off cemeteries. We cover boo-boos with Band-Aids. We are isolated from anything that connotes pain and death and in some regards we're isolating ourselves from reality when we do that."

A Natural Relationship with Animals

Nature, as others have argued, is not always a nice place. Animals in nature suffer. They suffer in illness. They suffer terribly in death, at the teeth and claws of a predator. And they suffer from starvation, if there aren't sufficient predators to keep the population in balance.

Foote says he has no argument with vegetarians and often goes a week or more without eating meat, especially if his freezer is empty of wild game. Foote makes clear divisions in the animal kingdom, between the game he hunts for meat, the livestock raised on farms and the companion animals he keeps.

Wild animals, he says, regard him as a predator. That is the natural relationship. But too many people, he says, mix up that relationship because of feelings they have for their pets.

"I think it's misplaced projection for people to look at wild animals and artificially project onto them these Disneyesque sentiments."

This transference of feelings from pets to wild animals is another symptom of urban disconnection, says Foote.

"You can poll people on the street and they just don't know what the real conditions are on a day to day basis 100 miles north of the city."

Van de Pitte agrees that humans mix up their feelings for animals. She says you can see it all over the world. If viewed on a global scale, the human relationship to food animals looks nothing if not hypocritical.

Cultural Differences in Food Choices

North Americans are horrified, for example, by the practice in Asia of eating dog. Horses, which are beloved here, are on the menu in some parts of Europe. Rabbit is a delicacy even domestically, but not one most North Americans can abide.

Hindus, meanwhile, are just as horrified by our practice of slaughtering and eating cattle, which they revere. Jews and Muslims don't eat pork.

"We have these arbitrary preferences in our culture for one thing over another as a food animal," says van de Pitte. "I think there is something really bizarre about refusing to eat bunnies and dogs and cats, but deciding to eat lambs and calves and other things."

If these are moral or spiritual decisions, as they are in most cases, should they be applied more broadly, to all animals, not just this one or that?

"I'm interested in arguments like: why do we have anti-cruelty legislation in the first place?" says van de Pitte. "It's a little bit inconsistent to have anti-cruelty legislation and then carry on with our regular practices."

Ultimately, we all have a responsibility to make up our minds, says van de Pitte.

Another philosopher, Plato, once said the unexamined life is not worth living. He meant that humans were obliged to think—to form beliefs and morals—not just to follow the herd, so to speak.

What might a personal animal-rights philosophy look like? Fortunately, free choice allows many options, from refusing to eat meat or wear leather, to strolling into a steakhouse in a fur coat, to somewhere in between.

Van de Pitte says, though, that a personal philosophy must consider morality. We are obliged, she says, to be consciously aware of animal rights issues. We must think about why we befriend some animals, while eating others. We have to consider whether their food preferences are just that—culturally prescribed preferences. Or is meat a practical necessity in our diet?

Deciding Whether to Eat Meat

Van de Pitte says we must also consider the costs of our decisions, not just to animals, but to the environment. Factory livestock operations are widely criticized as polluters.

"It all becomes, I believe, morally imperative that we think about it," she says.

"At the end of the day my final line is usually: do we have to use animals as we do? We are at a point culturally where we are capable of making choices.

"We are being urged to use them in traditional ways, partly from inertia and partly because these practices are now part of the global economy and some people make a lot of money from conducting business as usual."

But she wonders if many people wouldn't be happier if they related to animals differently, "recognizing that they are more like us than we ever wanted to suppose.

"We've always emphasized the differences, now we are recognizing the similarities. Who do we want to be? What kind of world do we want to live in?"

Foote fears that this kind of soul searching will go too far and that people will be shamed into giving up something they cherish. People should not be denied their right to make personal choices, he said.

"In many regards religion does that chronically, and the line between vegetarianism and religion is often crossed or confused," says Foote. "The same tactic, of projecting a moral right or wrong on other people, is used and it crosses the line of individual freedom.

"I think we have an obligation to have a lot of respect for contrary and contrasting views, and there's room in this country for all."

The ideal situation, says Foote, would be one where every person had to kill and process their own food animals. It would reduce meat consumption to a more appropriate level, he says.

"That's part of the responsibility chain," he says. "I don't like people washing their hands of the negative parts and just taking the filet mignon."

Trying a Meat-Free Meal

No such cut of meat is on the menu at Max's Light Cuisine, where tofu and vegetables rule, where people often line up outside the door to get a taste.

Max Pieroelie isn't a vegetarian, though he grew up eating very little meat. Nor is he religious about vegetables, he says.

Nowadays, he says, he's getting all kinds of people through the door: young, old, committed vegetarians, the curious. Many newcomers are cardiac patients, sent here by doctors to learn that you can eat healthy vegetarian cuisine and still enjoy the meal.

Pieroelie laughs when he tells the story of one regular group of diners he had on Sundays. It was a group of spiritual worshippers from a nearby congregation.

One Sunday they stopped coming and he never saw the group again. A few weeks later one of the members stopped by. She explained the absence was due to the complaints of some in the group. No, it wasn't the cooking or the service they disliked, she said.

"There's just not enough meat."

THE CASE FOR VEGANISM

Jeffrey M. Freedman

The choice to be vegan and forgo the use of animal products for any reason is a spiritual one for author Jeffrey M. Freedman. Freedman does not want to contribute to suffering in the world by exploiting the suffering of animals for his own use. He acknowledges that his personal choice will not dramatically lessen the violence and pain that occur in the world, but veganism is his way of choosing not to be a part of it. Having reverence for life, he argues, requires being conscious of one's actions and their impact on the world. Choosing veganism is one way of fulfilling the spiritual desire to decrease the suffering of living beings. Freedman is a writer and journalist who has contributed to publications such as *Tikkun* and *International Vegetarian Union*.

Being vegan is about more than what I do or do not eat. For me, it is a prayer, a petition asking why animals and people suffer greatly in a Universe created by a benevolent and loving God. This question led me to a lifestyle that is focused primarily on abstaining from the consumption or use of anything that comes from or contains animals or animal products.

Veganism is a corollary of *ahimsa*, the universal principle of compassionate, nonviolent living, the a priori maxim of Judeo-Christian ethics and Eastern spiritual philosophies. Mohandas Gandhi said "The greatness of a nation and its moral progress can be measured by the way its animals are treated."

Veganism, for me, is not so much about dietary abstinence as it is about spiritual sustenance; spiritual sustenance that fills the dark and empty spaces I feel lost in when I witness animal and human suffering, or anything that is an affront to what is Holy or good in the world. It is a lifestyle imperative that flows from my love of animals and reverence for life.

A Spiritual Decision

It wasn't until I got to university, on my own for the first time, that I realized there was a disconnect between what I felt in my heart (love for animals) and what I was putting in my body (corpses of animals), and that my spiritual life would have to mediate between and reconcile the two. It did. I stopped eating meat and chicken and—after I realized fish are not plants with gills—seafood went too. Becoming vegetarian made me feel I was doing something to lessen the suffering of animals (or that at least I wasn't contributing to it). I was making a statement about what my conscience could not live with and what my body could live without, but it also felt like an inadequate human response to a spiritual dilemma.

Why God's creation suffers, and how and when this suffering will cease is a question that has always tormented me—a mystery only God has the answer to. But I couldn't even bring myself to ask this question knowing that what I ate, what I wore, what I did, contributed to suffering in this world. For me, to be able to fully explore the question of suffering, I had to give up the products and by-products that can't be produced without causing suffering to animals—including meat, eggs, milk, fur, leather, wool, down, and cosmetics or chemical products tested on animals. (I include circuses, zoos, and all other institutions that confine or exploit animals in this list.) To the extent they don't cause suffering, in any way, I consider consuming, wearing or watching them acceptable.

Eradicating Suffering

This ethical standard I try to live by is predicated on Albert Schweitzer's "Reverence for Life," my desire to decrease, or at least not contribute to, the suffering of any sentient being; and the interrelatedness and common origin of all life on Earth. If she, he, or it suffers, I suffer. What constitutes suffering, as far as I understand it and the way most Buddhist's define it, is that everything/everyone wants to live and nothing/no one wants to feel pain. Anything that causes pain or death causes suffering. But eradicating all animal products from my diet, my clothes, and every aspect of my life is also, ironically, a statement about my powerlessness in the face of the world's suffering. I have had to admit that what I don't eat isn't going to have a major impact on the violence and the suffering of the innocents in the world; that it would take more than my abstinence from eating animals to bring about a state of ahimsa to the world. Veganism, for me, is asking God to do what I am incapable of doing myself.

I read something in the news recently about the ongoing abduction and breaking of baby elephants in Thailand. They are taken from their mothers, tied by their feet so they can't move, beaten with sharp instruments on their head till they bleed, and kept awake by loud noise, sometimes for days. This torture goes on until they either go mad or become docile enough to perform in circuses and tourist attractions.

Two blocks from where I live and work an injured pigeon has been cowering under a store ledge trying to avoid the prowling cats, blinding snow, wind, and other urban predators. Hundreds of people have passed by and ignored him the way they ignored the mangled pigeon I found during one of last summer's most unbearably hot and humid days. He was attacked by a cat, couldn't fly, hobbled on one leg, looked unbearably sad and worn out. When I take these animals to the local wildlife rehabilitation center I am as much pained by the broken-hearted indifference of the other people who saw their suffering and did nothing as I am by the suffering itself.

For anyone sensitive to the suffering of animals and people who cannot defend or fend for themselves, these are the things that rend the heart and are a call to action and prayer. They are a call to action because to do nothing is to court helplessness and depression and defeat. They are a call to prayer because in an imperfect world, suffering, which is a symptom of separation from the Divine or the whole of creation, must exist. Prayer then becomes the last refuge of those who suffer greatly as a result of bearing witness to great suffering.

Reverence for Life

I realize this is somewhat of a spiritual-evolutionary leap; that I am, to a certain extent, intervening in the process of natural selection and survival of the fittest in my desire to pre-empt or lessen the severity of suffering in the 'animal kingdom.' People are always reminding me that, in nature, the big fish eat the little fish, we all prey on and consume some thing or someone, that it's a dog-eat-dog world. I know all that. I know it. But a fundamental principle of evolution is that those who adapt to the environment most efficiently increase their odds of survival. The predatory, polluting, war-mongering behavior of humanity has pushed us—and unfortunately most other life forms—perilously close to extinction. So is it too much of a stretch to posit that reverence for life, eating as low down on the food chain as possible and a desire to preserve that which sustains life (e.g. lakes, rivers, streams, forests, the ozone), may be a last-ditch attempt by the evolutionary, self-preserving wiring in us, if not the entire life-force of the planet, to move us further from the precipice of extinction?

We have polluted, consumed, caged, corrupted, deracinated, tortured, and tormented just about every form of creation on Earth. I think ahimsa and veganism are both a symbolic and a very real way to reverse this trend. Disregarding the sanctity of life and the planet with impunity has given us a world that is rife with violence, war, pestilence, starvation and poison in our air, water, and food. Being vegan is a way of becoming conscious of our actions and their consequences. It is a way of giving thought to the pain, suffering, and terror that occurs as a result of turning a living, breathing, feeling animal into the meat we put inside ourselves or in front of our children; a way of considering the global consequences of deracinating another forest, poisoning another river, depleting more of the ozone, feeding cattle on arable land that could be used to eradicate world hunger.

Like a fast at Yom Kippur or Christian Lent, I am trying to make myself ready to petition God to rid the world of suffering and violence that I can't personally eradicate or change in my lasting or globally significant way. I am asking Him to do something about the baby elephants and the wounded pigeons and the broken hearts of the world.

CHAPTER 4

IS ANIMAL
EXPERIMENTATION
ETHICAL?

Contemporary Issues
Companion

A HISTORY OF ANIMAL EXPERIMENTATION

Vaughan Monamy

In the following selection, professor Vaughan Monamy reviews the history of using animals in experimental research. Monamy relates that animals have been used in experiments since ancient times. Historically, Monamy explains, humans have viewed animals as objects lacking souls and the ability to reason. This philosophy justified the use of animals for experimentation. As biomedical studies using animals have continued into modern times, Monamy writes, methods have been refined, allowing important discoveries to be made. While Monamy emphasizes that animal experiments have resulted in important advances for medicine and science, he acknowledges that people may choose to find animal experimentation ethically unacceptable. Monamy lectures in environmental science and environmental ethics at Australian Catholic University in Sydney.

Early records of vivisection procedures provide sobering reading. However, it is worthwhile to examine some of them in order to understand how public concern over animal experimentation arose. We need also to consider the origins of western scientific practices and the prevailing societal attitudes towards them. . . .

Live animals, both human and non-human, appear to have been first used in ancient times principally to satisfy anatomical curiosity. In the third century B.C., the Alexandrian physicians Herophilus and Erisistratus are recorded as having examined functional differences between sensory nerves, motor nerves and tendons. Galen of Pergamum (129–199 A.D.), a Greek physician working in Rome, catalogued these early experiments, as well as conducting his own. He described, for the first time, the complexities of the cardiopulmonary system, and speculated on brain and spinal cord function. All such procedures were conducted without anaesthetics (which were not discovered until the mid-nineteenth century) and it is interesting to note the expression of his feelings during such experiments. When investigating the anatomy of the brain, Galen preferred to vivisect pigs to

Vaughan Monamy, *Animal Experimentation: A Guide to the Issues*. New York: Cambridge University Press, 2000. Copyright © 2000 by Cambridge University Press. Reproduced by permission of Cambridge University Press.

'. . . avoid seeing the unpleasant expression of the ape . . .' Galen left a legacy for future scientists. In *De Anatomicis Administrationibus* (On Anatomical Procedures), he detailed precise experimental methods and indicated which instruments would be best to perform many specific procedures.

Documentation of vivisection from the Dark Ages is scanty. It was not until Galen's records were re-discovered during the sixteenth century that there appears to have been any renewal of interest in anatomy and scientific methods. Such experiments often were conducted as public demonstrations. Belgian Andreas Vesalius (1514–64) and his students in Padua, Italy, illustrated public lectures on anatomy by using systematic non-human vivisection. An animal, usually a dog, would be cut open while still alive and the function of each organ would be speculated upon as it was located. It appears, from the records of these procedures, that the welfare of their experimental subjects was a low priority for these early vivisectionists. Maehle and Tröhler recorded that the experiments of one of Vesalius' pupils, Realdo Colombo (1516–59), involving pregnant dogs, were greatly admired by members of the Catholic clergy:

> Colombo pulled a foetus out of the dog's womb and, hurting the young in front of the bitch's eyes, he provoked the latter's furious barking. But as soon as he held the puppy to the bitch's mouth, the dog started licking it tenderly, being obviously more concerned about the pain of its offspring than about its own suffering. When something other than the puppy was held in front of it's mouth, the bitch snapped at it in a rage. The clergymen expressed their pleasure in observing this striking example of motherly love even in the 'brute creation'.

The Christian View

It may be difficult for readers to understand the apparent indifference to suffering exhibited in southern Europe at this time. What must be considered, however, is that the Christian church subscribed to the view that humans, blessed with the divine gift of reason, did not share a common lineage with other animals. Three hundred years earlier, St Thomas Aquinas (1225–74) had declared in his *Summa Theologiae* (1260) that humans were unique; all other animals were incapable of rationality because they possessed no mind. Only humans had a soul and the power to reason. Without a soul, non-human animals were merely objects, devoid of personality or rights. They existed only for human needs and were bereft of moral status. This is not to say that the Christian church supported a view that an absence of moral status meant that any form of cruelty was acceptable. The church recognised that the animals over which humans had been given dominion were a part of God's creation and, for that reason, were worthy of respect. Many animals, such as the dove, were symbolised as a part of Christian worship, and St Francis of Assisi was venerated because of his sympathetic attitudes towards animals. At the same time, however, Christian society did not see the infliction of pain on animals (or humans for that matter) as objectionable in itself, if it was an unintended consequence of some 'higher' purpose. However, the gratuitous infliction of pain was viewed as morally reprehensible cruelty. The inescapable suffering of animals during experimental procedures, such as that described above, was not seen in any way as cruel while it was conducted in the pursuit of greater knowledge.

The Influence of Descartes

The seventeenth century saw an explosion of interest in scientific activity. British Lord Chancellor Francis Bacon (1561–1626) sustained the Christian anthropocentric (human-centred) view in his *De Augmentis Scientarium* (The Advancement of Learning.) He asserted that much could be learned of the human body and its workings by vivisecting non-human animals and that such dissection obviated the need for the morally repugnant (but nonetheless fairly common) practice of human vivisection involving criminals.

Philosopher René Descartes (1595–1650) was to play an important role in early debate over vivisection. Christian-centred humanist attitudes, so prevalent throughout Europe, became exaggerated into a mechanistic philosophy following the publication of Descartes' *Discours de la Méthode*. Here, Descartes stated that it was possible to describe humans and other animals as complex machines: their bodies would obey known laws of mechanics. Descartes also believed, however, that the divine gift of the soul distinguished the human animal from all others. Only humans were conscious and capable of rational thought. Only humans were capable of acts of free will, and had true language. Only humans could declare *Cogito ergo sum*—'I think therefore I am'. The reactions of non-human beings were dismissed as mere reflex, the response of automata. This concept of 'beast-machine' was critical to the way in which scientists viewed other animals. It provided a convenient ideology for early vivisectionists: how could animals suffer real pain if none had a soul? How could animals suffer real pain if none had real consciousness? In Descartes' writings was found a reason to discount the behavioural responses of animals to vivisection (which would be described as symptomatic of pain in humans) as the mere mechanical reactions of robots. Cries of pain in non-humans were now interpreted as the squeaking of un-oiled cogs. . . .

The Rise of Modern Biomedical Studies

In a series of formative experiments conducted at the anatomy school in Padua in 1628, Briton William Harvey (1578–1657) demonstrated the circulation of blood using animals, extrapolated the discovery to humans and in so doing showed the value of vivisection not only for satisfying *anatomical* curiosity, but also for comparative *physiological* investigation. Questions, long pondered, about how we breathe, digest food and so on suddenly appeared to have physiological answers. As a result of Harvey's experiments, many other scientific investigators were eager to delve into the workings of the animal body. The rate of animal experimentation increased—an increase that was to continue beyond the seventeenth and into the eighteenth and nineteenth centuries.

Frenchman François Magendie (1783–1855) was among the first to determine that many bodily processes resulted from the co-functioning of several organs. This realisation set in train numerous experiments that involved manipulative procedures rather than just internal observations. Although many of his experiments were 'hit-or-miss', Magendie is described as the founder of modern physiology.

A Rationale for Experimental Medicine

Another landmark in physiology came with the publication of *Introduction à l'Etude de la Médicine Expérimentale* by one of Magendie's students, Claude Bernard (1813–78). In this work, Bernard declared that a precise approach to experimentation must involve the study of one parameter while holding extraneous variables constant (this remains as a fundamental approach in modern science). In addition, he responded to a growing number of critics of vivisection by offering a powerful philosophical rationale for experimental medicine. Bernard posed:

> Have we the right to make experiments on animals and vivisect them?. . .I think we have this right, wholly and absolutely. It would be strange indeed if we recognised man's right to make use of animals in every walk of life, for domestic service, for food, and then forbade him to make use of them for his own instruction in one of the sciences most useful to humanity. No hesitation is possible; the science of life can be established only through experiment, and we can save living beings from death only after sacrificing others. Experiments must be made either on man or on animals. Now I think that physicians already make too many dangerous experiments on man, before carefully studying them on animals. I do not admit that it is moral to try more or less dangerous or active remedies on patients in hospitals, without first experimenting with them on dogs; for I shall prove . . . that results obtained on animals may all be conclusive for man when we know how to experiment properly.

The work of physiologists such as Magendie and Bernard, coupled with the discovery of the anaesthetic properties of ether (by Crawford Long in 1842, and by William Morton in 1847), resulted in an adoption of technically sophisticated surgical procedures. Animal experimentation became routine in an increasing number of physiology laboratories throughout Europe. In Britain, the 1876 *Cruelty to Animals Act* required meticulous registration of the numbers of research animals used in experiments each year. These records show that the number of procedures involving research animals increased from 311 in 1880 to over 95,000 in 1910.

Medical Breakthroughs

The end of the nineteenth century saw vast improvements in aseptic surgical techniques and the development of bacteriology and immunology. Key medical discoveries, such as the discovery, in 1882, of the bacterium responsible for tuberculosis, and of a diphtheria antitoxin in 1894 (which rapidly reduced infant mortality from 40 per cent to 10 per cent in those afflicted), led to broad public acceptance of animal experimentation.

More medical breakthroughs occurred at the beginning of the twentieth century, further emphasising the value of using animals in biomedical research. These included: the extraction of the first hormone (1902); a chemical treatment for syphilis (1909); and the isolation, by Banting and Best (1920), of insulin, leading to the development of an effective treatment of diabetes mellitis. Such spectacular advances attracted enormous public acclaim and heralded the modern era of animal experimentation. In Britain, the numbers of animals used in experiments increased to exceed one million per year in 1943 and five million per year by the mid-1970s. Numbers had declined to three million by 1991, and were down to 2.7 million in 1997.

Increased government financial support led to the important improvements in preventative medicine and surgical techniques that today permit many to enjoy longer and enhanced lives. In 1989, the American Medical Association Council on Scientific Affairs published an impressive list of medical advances made possible through research using animals. It included studies of autoimmune deficiency syndrome (AIDS) and autoimmune diseases, anaesthesia, behaviour, cardiovascular disease, cholera, diabetes, gastrointestinal surgery, genetics, haemophilia, hepatitis, infant health, infection, malaria, muscular dystrophy, nutrition, ophthalmology, organ transplantation, Parkinson's disease, prevention of rabies, radiobiology, reproductive biology, shock, the skeletal system and treatment of spinal injuries, toxoplasmosis, yellow fever and virology. Such research has resulted in enormous gains in human knowledge, with subsequent benefits for human and non-human health.

The Benefits of Animal Experimentation

This is an important point that deserves emphasis. We live in an unprecedented age in which life-threatening illnesses are kept at bay to an extraordinary degree. Having lived all of our lives at such a time, it is easy to forget that as recently as 50 years ago many diseases, such as polio and tuberculosis, were common killers in our society. In early Victorian Britain, life expectancy at birth was 42 years. Today, life expectancy at birth in most western nations exceeds 70 years. One important reason for this increase in longevity (without detracting from, for example, the role of enhanced public health measures, clean water and occupational safety laws) is the benefits that have stemmed from animal experimentation.

Given such a track record, how could anyone condemn such practices? Surely increases in human health standards, as well as increased productivity of domestic livestock or increased general knowledge of wild animals through zoological and ecological investigations etc. outweigh any suffering involved in obtaining these advances?

This is at the heart of the matter. Some see that all experimentation is vital, ultimately beneficial and must be allowed to continue unchecked. At the other end of the spectrum are individuals who hold deep convictions that all animal experimentation is an abuse of other species for selfish human gain. If you choose a subset of humanity (say, the readers of this book) and quiz them on their personal attitudes, all will opt for a position somewhere along this continuum. Where you choose to stand will depend on many things, including career aspirations, vested interests, level of understanding of complex issues, personal moral views, religious beliefs, and levels of compassion for certain other animals.

ANIMAL EXPERIMENTATION IS NECESSARY

Carl Cohen

Carl Cohen is a professor of philosophy at the University of Michigan in Ann Arbor. In this selection, Cohen contends that experiments on animals are necessary in order to safeguard and improve human health and well-being. Vaccines for diseases such as polio and malaria could not have been developed without animal testing, Cohen explains. Critics of animal experimentation argue for developing alternatives such as the use of tissue samples and computer simulations, but Cohen asserts that these alternatives will not yield appropriate data on how new medicines will affect humans. In addition, the author continues, opponents of animal experimentation should remember that most laboratory animals are rodents that are bred for biomedical use, and human health needs morally justify the use of these animals in research.

In the summer of 1948, I was a counselor in a children's camp in western North Carolina; one of our young campers was stricken by polio, an event that befell summer camps around the country that year. The fear in the faces of parents who rushed to the camp to take their children from that dangerous place, reaching with their arms across the quarantine chains, is vivid in my recollection. They trembled; my parents trembled; all parents trembled every polio season, which was every summer, never knowing whether their children would prove to be among the random victims of that crippling and often fatal disease.

The epidemic did not soon relent. During the summer of 1952, more than fifty-eight thousand American youngsters contracted polio. Thousands of these children died; thousands more were sentenced to a lifetime in the cruel machine that we called "the iron lung."

A vaccine for polio had been under development for some time, but years would pass before its safety and reliability were established and it could become widely effective. By the late fifties, however, childhood vaccination for polio had become routine; by the close of

Carl Cohen and Tom Regan, *The Animal Rights Debate.* Lanham, MD: Rowman & Littlefield Publishers, Inc., 2001. Copyright © 2001 by Rowman & Littlefield Publishers, Inc. All rights reserved.

that decade, the number of reported polio cases in the United States had been reduced to twelve—*one dozen*. Polio was totally eradicated from the Western Hemisphere not long after that, and as I write the disease is nearing eradication in other parts of the world as well.

The astounding success of that first polio vaccine was announced at the medical center of the University of Michigan, only blocks from where I live in Ann Arbor. Its impact has been global. How many have been spared misery and death by this one great step in medical science we can hardly guess. But about this wonderful vaccine and its successors we do know one thing for certain: *It could not have been achieved without the use of laboratory animals.*

To prepare the culture from which the polio vaccine was made, animal tissue was indispensable. And with that new vaccine greatest caution was obligatory. Many candidate vaccines had earlier been tried and had failed. From those earlier vaccines some healthy children had actually contracted polio. That could not be allowed to happen again. To test the new vaccine before its administration to humans, animal subjects were absolutely essential.

Animal Experiments Are Critical for Human Health

This true story, close to us in time and place, is widely known. But there are a thousand stories like it of which we are mostly unaware: scientific victories over tuberculosis and typhus, the discovery of insulin rescuing diabetics from misery and death, the discovery of antibiotics and the development of anesthetics—uncountable advances that have proved to be of incalculable importance to human beings. All this and much more could not have been done without the use of animals in the key experiments. The absolutely critical role of animals in these investigations cannot be emphasized too strongly. It is not simply that animals were convenient in such work or that they speeded the results—although if only that were the case, the justification of their use would be strong enough, heaven knows. But it is not like that. Experiments using animals are not merely helpful; they were and remain a necessary condition for most critical advances in protecting human health.

New vaccines are always dangerous. Testing them (and testing many new drugs as well) unavoidably risks the well-being, sometimes even the lives, of the first experimental subjects. When a new vaccine or new pharmaceutical compound has been devised and is at last to be tried, whose lives shall we put at risk? Not the lives of my children, certainly. The lives of yours, perhaps? You are offended by the thought—rightly. Shall we then use the children of unsophisticated people in underdeveloped societies? Heaven forbid. What, then, are our alternatives? There are only two: We will use animals (by which I mean nonhuman animals, of course) in such biomedical experiments, or we will not do those experiments at all.

The philosophical dispute about animal use therefore concerns each of us directly, insofar as we ourselves use, or expect to have available for use, vaccines and drugs that are effective and safe. Respecting every vaccine and almost every new drug, we will use animals in the research process, or we will not develop and we will not have that vaccine or that drug for our use.

There Is No Good Alternative to Animal Experiments

Much talk in recent years suggests that animals in biomedical experiments should be replaced. Let us use "tissue samples" instead of whole animals, it is said; let us use "computer simulations" of the disease or the experimental organism. In a very few restricted settings, such replacements are possible and appropriate. But in most medical investigations the replacement of live animals with tissue samples or computer simulation is simply out of the question. The reason is simple: Investigators must learn the impact of a new compound or a new vaccine on the living organism as a whole; side effects that may be dangerous can be investigated only in the living organism and its complicated network of constituent organs, as they actually function. Computers cannot give that information. The results of experiments using tissue samples, however favorable, will not be enough to warrant the use of a new drug in humans until we have done our very best to learn its full organic impact. We can learn that only by studying the outcome of its use on live animals who are not human.

The first use of a new compound on a living organism is inescapably experimental. The subject of that experiment will be a human or another animal. The use of humans in such experiments we will not permit, understandably. If, therefore, the use of nonhuman animals is also not permitted, there will be no such experiments.

The large-scale replacement of animals by tissue samples or computers or anything else is, to be blunt, a misleading fantasy. In this continuing discussion of the morality of animal use, all such wishful conjectures should be put aside. Where alternatives to the use of animals can yield the needed data, it is right to use them, and it is right to seek such alternatives, as we do. But romantic dreams cannot guide actual research and may prove dangerous if relied on. Professor Tom Regan (who condemns all animal experimentation as immoral) was asked publicly in Washington, D.C., what he would have medical investigators do if the use of animals were indeed forbidden. How should they proceed? His answer was that they must find alternatives somehow, somewhere. They must, Regan said, "set their imaginations on fire." That is an embarrassing response, not very helpful advice from a philosopher to laboratory scientists. To those who labor for new treatments of stubborn diseases, and for those who suffer grave illness and who pray for relief and can only hope for the successful outcomes of investigations in which experimental animals are the key, "set your imaginations on fire" is close to insult. We use animals because in most medical contexts no known alternatives to animal use are available. It is virtually certain that no such alternatives will exist for a very long time at least, since none are even on the horizon. In fact, it is probable that there never will be feasible alternatives to the uses of animals in much of medicine.

Mice and a Malaria Vaccine

The killer disease for which a vaccine is now most desperately needed is malaria, which infects about *three hundred million new victims each year*, of whom *more than two million die* every year, most of them children in Africa and Asia. Drugs to combat malaria have become less and less effective as new strains of the parasite, resistant to those drugs, arise and spread. In the United States also malaria is spreading, the number of cases in recent years up well more than 200 percent—attributable to people who return from visits in their home countries, where they find that they are not immune. Many vaccines have been tried—not recently on children, thank God—but have failed. The parasite that causes malaria, carried by the anopheles mosquito, is so resilient that scientists have long been unable to generate in any living organism the antibodies needed to ward off the disease.

Recently, however, some striking progress has been made in this battle. After decades of effort a vaccine has been developed and tested at the Naval Medical Research Institute in Bethesda, Maryland; it does inoculate with safety and complete success against malaria—*in mice*. For humans we do not yet have a vaccine proven safe. But before very long we probably will have one, and we will get it, if we do, only because we were able to experiment on mice, many mice, who will have been deliberately killed by investigators to learn what must be learned for the development of that new vaccine. In developing a malaria vaccine, we will use mice (as we do in the study of cancer, and diabetes, and hundreds of other human diseases) *because there is no other way.*

Most Experiments Are Done on Rodents

Many wince at the thought of using animals in biomedical research because they think immediately of dogs and cats, whom they love. In this view we are misled. The controversy should be understood to be one that mainly concerns the use of rodents, and among rodents chiefly mice and rats. Dogs, pigs, and other mammals (almost always anesthetized) are also used when they are the most suitable models for the disease under investigation. But only in a small minority of studies is that the case. The number of dogs and cats killed each year as experimental subjects is less than one-fiftieth of the number of dogs and cats killed *in animal shelters by humane societies for conve-*

nience, because we have no place for them. About ten million dogs and cats are put to death in the United States each year for no good reason save that nobody wants them. Bearing in mind this wholesale killing of strays and former pets in animal shelters, how ought we respond to academic philosophers like Tom Regan who strenuously protest the carefully limited use of mice by medical science? Of all the animals used in biomedicine, dogs and cats make up less than 1 percent and primates less than three-tenths of 1 percent. Pigs, rabbits, and chickens are used more—but they amount to an extremely tiny fraction of 1 percent of all those billions of pigs, rabbits, and chickens killed for use as human food.

The U.S. Department of Agriculture recently estimated the number of animals used in medical and pharmaceutical research to be about 1.6 million, of which the vast majority, approximately 90 percent, were rats, mice, and other rodents. These animals would not have come to exist had they not been bred specifically for biomedical use.

Rodents Are a Threat to Human Health

Meanwhile, in the world of everyday life outside science, the extermination of rats and other vermin that infest our cities is a perennial objective, difficult to achieve but important for the sake of human health. The rats that multiply in our central cities are dirty and dangerous animals, carriers of disease, and specially threatening to the poor. In Chicago, where until recently rats outnumbered people by more than two to one, an aggressive campaign to clean the lakefront of rats has had substantial success. Of the rat population of about six million in 1979, more than five million had been eliminated by 1997.

In Boston, a massive rat control enterprise, largely successful, has recently been made necessary by the Central Artery Tunnel project, called by locals "the big dig." A *huge* population of rodents, whose ancestors have been burrowing and breeding in the vicinity of the old Boston harbor for centuries, might have been dislodged by the construction of the new tunnel/highway and forced into the central city. These sewer rats (*Rattus norvegicus*) are not the cute little pets to be found in preschool classrooms; they are large (over a foot long), ferocious, often filthy creatures laden with disease, rats that eat virtually anything—including human babies when given the opportunity. The feces and urine of wild mice (*Peromyscus maniculatus*, not these rats) have very recently transmitted the deadly hantavirus, resulting in many human deaths in the United States. Rats like those of concern in Boston transmitted bubonic plague, the "black death," in ages past. Should those Boston rats have been protected, possibly chased into the alleys and basements of the crowded city center? Or should they have been poisoned, as they were? In my judgment, it would have been morally wrong to risk the invasion of the human habitations of Boston's poor by these rodents; it was right to kill them as humanely and as efficiently as possible. Readers are likely to judge similarly about any rats, possibly disease-laden, discovered in their own basements.

Safeguarding Human Health

What, then, will be our considered view of the protection we owe to rats on moral grounds? This is not merely an abstract puzzle. Deciding what conduct is moral is, as Kant insisted, a very *practical* matter. What shall be our practice, our actions, in dealing with rodents? Do we seriously think it to be wrong to kill disease-carrying rats for the sake of human health? Wrong to use rats as the subjects of experiments designed to develop new drugs for cancer, new vaccines for malaria and other human afflictions?

Whether we are morally justified in using animals as we do in science must be decided in the light of what we know about safeguarding the healthy and curing the sick. The arguments in the continuing philosophical controversy should be evaluated in the light of the facts, often very unpleasant facts, of human and animal disease.

ANIMAL EXPERIMENTATION IS UNNECESSARY

Sarah Rose A. Miller

Sarah Rose A. Miller was a fifteen-year-old high school student when she wrote this prize-winning essay for the *Humanist*, a magazine published by the American Humanist Association. Miller argues that animal experimentation is cruel, expensive, and unnecessary. Not only do animal experiments subject animals to pain and distress, she writes, the results of the experiments often lack validity. In addition, Miller reports, huge amounts of money are spent on experiments whose results are pointless or repetitive—money that could be better spent on helping improve people's lives. Finally, Miller asserts that just because people can experiment on animals does not mean they should. She points out that there are many viable alternatives to animal experimentation, and these should be explored as a first step to achieving humane treatment for animals.

"Every second of every day of every year, an animal dies in an experiment in the United States," reports Last Chance for Animals. All in the name of science. Scientists cut open live animals to see how their bodies work, they poison animals to see how their bodies react, they test cosmetics on them, and more. These animals often die excruciatingly painful deaths after having lived their entire lives in isolation, trapped in tiny cages with minimal stimuli.

On the other hand, millions of people would die each year from cancer, polio, diabetes, heart disease, and kidney failure if research on nonhuman animals hadn't been practiced since way back when. Animal research has been an important part of many biomedical breakthroughs, allowing scientists to test on animals countless possible cures for these illnesses.

Many people are unsure with which side of the argument they agree in regard to this issue, because millions of lives are lost either way. Although more animals die (and suffer much more before they die), humans are our species; if faced with the choice of saving a random animal or a random child, most people would save the child.

They can relate more with children, knowing that children feel pain and distress, while it wasn't a commonly accepted fact until recently that animals have feelings as well.

Animals are used for research in several different ways: for education, for testing cosmetics, and for biomedical research. In regard to education, children all over the world dissect animals to learn their basic anatomy. It gives them a chance to see the different organs and their placement and to imagine what the creature's bodily systems were like when it was alive. It also leaves a more lasting impression than if the children merely study from a diagram. But it also leaves the impression that animal lives can be wasted if this benefits humanity. Millions of animals, mostly frogs, are killed every year expressly for educational use. Sometimes, in the act of removing thousands of a particular species from an area, the biosphere where they live is destroyed or dies because it is no longer supported by a specific species. This shows no respect for the lives of other species, only self-interest and the sense of self-gain.

Testing Products on Animals

Animals are used for product testing to make sure that a certain product can be safely marketed without the risk of causing serious injuries. From hair conditioner to pharmaceuticals, lab animals have products sprayed on them, put in their eyes, rubbed into their skin, or given to them to consume or inhale.

One commonly used procedure is the Draize Test, which is used to test household products for harmful chemicals. In the Draize Eye Irritancy Test, solutions are applied immediately to the subjects' eyes—usually a group of albino rabbits. This often causes intense pain and destroys the rabbits' eyes, leaving them raw and bleeding. After the test period (around seven hours), the rabbits are killed so they can be examined for internal damage.

Another standard test is the LD-50 (Lethal Dose 50 Percent) Test. This procedure is used to measure the toxicity of a substance—the amount of the substance it takes to kill half a group (generally 200) of test animals. The test usually goes on for days, and the animals suffer extreme pain and distress. When it's over, the animals remaining alive are killed. The Draize and LD-50 tests are only two of numerous types of product tests commonly administered to animals.

Animals and Biomedical Research

The use of animals for biomedical research is the most controversial of the three previously mentioned ways animals are used in research. The prevalent methodological use of animals in biomedical research is vivisection—the act of cutting into a creature still alive. This is done mainly to gain knowledge of how the bodily systems work and to examine how they react to various stresses and viruses. The numerous other experiments to which lab animals are subjected are sometimes placed under the heading of vivisection as well. These include: burning live animals (so that severely burned tissue may be examined); inflicting hard blows to the head, deep cuts, and other serious injuries to see how their bodies will react; injecting them with fatal viruses like HIV; poisoning them; and so forth. These experiments help scientists and physicians find treatments so humans will not have to suffer so much—so that thousands of children won't get polio, so that diabetes won't kill every person who acquires it, so that cancer can be treated and perhaps cured, so that a variety of illnesses may be fought.

The main problems with animal research are the validity of the test results, the expenditure, and the morality of it. The physiological difference between humans and animals exists—even with chimpanzees, our closest relatives—and consequently many test results come out incorrect. Innumerable people get seriously ill, develop liver or heart damage, because the prescribed drugs had unpredicted "side effects." According to physicians Neal D. Barnard and Stephen R. Kaufman:

> The U.S. General Accounting Office reviewed 198 of the 209 new drugs marketed between 1976 and 1985 and found that 52 percent had "serious postapproval risks" not predicted by animal test or limited human trials. These risks were defined by adverse reactions that could lead to hospitalization, disability, or death.

The mental state of the lab animals can also produce misleading results. If the animals are in a state of extreme distress from being trapped in tiny cages their whole lives with no stimuli and no room to move around, it can induce their various bodily systems to behave differently.

Wasting Money on Animal Research

Animal research is quite expensive: $18 billion is spent annually by vivisection industries in the United States. Some of it funds research programs that actually accomplish something useful, but often enough this tax money is completely wasted on duplicated or pointless experiments. And even if an experiment does have a purpose, that much money doesn't generally have to be spent on it. For example, the National Institutes of Health spent:

• $1,329,332 on project PO1 HD2253900-01, conducted by Boston University, to demonstrate that malnourished rats bear offspring which are mentally retarded.

• $772,963 on project PO1 ES04766-08, conducted by Oregon State University, to test carcinogens in rainbow trout as possible indicators for humans, as well as testing current applications of human chemotherapy on rainbow trout.

• $969,475 on project R35 CA49751-10, conducted by Baylor College of Medicine, to stimulate glandular secretions in female rats in order to observe them engage in sexual activity. Instead, any extra money could be directed toward making the lives of the lab animals a little better by providing them with larger cages. In her book *Through a Window*, Jane Goodall said:

> When people complain about the cost of introducing humane living conditions [to the lab animals], my response is: "Look at your lifestyle, your house, your car, your clothes. Think of the administrative building in which you work, your salary, your expenses, the holidays you take. And, after thinking about those things, then tell me that we should begrudge the extra dollars spent in making a little less grim the lives of the animals used to reduce human suffering."

Too much money is wasted on pointless experiments while millions of people in the United States can't afford to go to the hospital, can't afford a good education, can't afford food and are malnourished, aren't able to do what they would like with their lives because they can't risk losing a job, and can't live happy lives because they are constantly worrying about money. Millions of people are under this kind of stress daily.

A Question of Morality

Finally, the morality of animal research must be addressed. The primary reason people have challenged the use of non human animals in scientific research is because of one question: do we have the right to cause other animals to suffer so that we don't? Many centuries of intense rivalry are behind this question.

Is it fair that we cause other beings intense pain and distress because we can? Lab animals live their lives in isolation. They usually have nothing to do except sit in their cages and stare at the walls, maybe rock back and forth to make it more interesting, day after day. Eventually, most animals will have reduced brain activity. They will be like lobotomy patients. On the day when something different happens—when they are injected with a fatal virus or taken out of their cage to be experimented upon—they are probably very bewildered. Something is going to change, the endless monotony of life is going to cease, but is it a welcome or unwelcome change?

Perhaps many of the animals would prefer to be dead, even if it meant going through excruciating pain—perhaps not. But they don't know what is happening to them when they are being experimented on; they don't know why this is happening to them. They only know that they are in extreme pain at the moment, and the moment is all that matters to them. And they know that humans are doing this to them—the same humans who have kept them isolated to live their entire lives in a monotonous vacuum.

Finding Alternatives

There are alternatives to animal testing, but most scientists and physicians don't seriously consider using them instead of animals.

Animal research is morally wrong, but it has been useful in the past. So perhaps the answer is to take everything that was learned from past research and how the body functions and apply this to newer technologies like advanced computer simulations. We could accept the wrongs committed to animals in the past and turn them into rights, helping human lives without sacrificing animals.

We must remember what has happened in the past, like with the Nazis. They considered Jews to be beneath them—to be animals with no feelings—and so used them for experiments. Nazis justified their actions by saying that they were aiding the superior race and, besides, Jews didn't really count as anything important. That is exactly how most humans look upon other species today. But we don't think it is all right to treat other human beings this way, no matter what race, so how can we think it right to do the same thing to other species? Jane Goodall put it very well when she asked the question: "How can we, the citizens of civilized, Western countries, tolerate laboratories which—from the point of view of the inmates—are not unlike concentration camps?"

We must change that. Even if it takes time, we must change the fact that we are treating other species of beings that live on this planet as the Nazis treated the Jews during World War II. We must fight to get animals more humane treatment, and we must start using the quicker, easier, less expensive, and much more reliable alternatives to animal research.

ALTERNATIVES TO ANIMAL EXPERIMENTATION WILL MAKE IT OBSOLETE

Andy Coghlan

Animal experiments may soon be a thing of the past, writes Andy Coghlan in the following selection. The Organisation for Economic Cooperation and Development (OECD), an international group that includes the world's wealthiest nations, has approved alternatives for testing chemical safety that do not rely on animals, Coghlan writes. This step may be just the beginning for widespread implementation of alternative techniques, he explains. Coghlan reveals that alternatives to animal testing are becoming the norm in Europe. In spite of resistance from the Environmental Protection Agency, the United States is likely to adopt them as well. Coghlan writes for *New Scientist* magazine.

Several million rabbits, mice and rats are destined to die in the name of science over the next 10 years. To some people, this is an unforgivable slaughter. To others, animal testing is the only adequate way to test new treatments for diseases and ensure that drugs and other new chemicals are safe.

But [since 2002], governments have been quietly signing deals that could dramatically affect this highly emotional and polarised debate. There are signs that those in power are finally paving the way for a global movement that could one day consign experiments on live animals to history.

The impetus has come from an unexpected source: the Organisation for Economic Cooperation and Development (OECD), the private "club" of the world's richest nations. In May [2002], OECD members agreed for the first time to approve four new tests for chemical safety that don't rely on live rabbits and rodents. These four tests alone could spare millions of animals each year from undergoing lab procedures.

The decision is a triumph for the OECD, which some critics say should shoulder most of the blame for the slow progress in introduc-

Andy Coghlan, "Vive les Animaux," *New Scientist*, vol. 175, July 20, 2002, p. 14. Copyright © 2002 by Reed Elsevier Business Publishing, Ltd. Reproduced by permission.

ing alternatives to animal testing. The organisation's strict voting system means that every member has to agree to a test before they can all start using it as a recognised standard.

"At the OECD, if one person won't play ball, discussion stops," says Michael Balls, outgoing chairman of the European Union's European Centre for the Validation of Alternative Methods (ECVAM). In an ECVAM report [that came out in July 2002], Balls condemns the slow rate of progress. In Britain, a report from the House of Lords is also expected to call for more research into alternatives.

But while other campaigners agree that replacements for animal tests aren't appearing fast enough, they say the recent decisions by the OECD signal a new era in establishment attitudes. "The first [new tests] going through the OECD are setting a precedent that will build confidence in the alternative techniques," says Gill Langley, scientific adviser to the Dr Hadwen Trust for Humane Research, a charity backing replacement of lab animals. "I hope that as these new tests come on-stream, new methods will be accepted more quickly."

So what's been holding things back? One country must bear particular responsibility for the OECD's past lethargy says John McArdle, director of the Alternatives Research and Development Foundation in Apple Valley, Minnesota. He points the finger at the US and its Environmental Protection Agency (EPA), the federal body charged with protecting American citizens and the environment from hazardous chemicals.

Waiting for the EPA

Consider the notorious "lethal dose 50" (LD50) test for acute toxicity. Groups of animals receive successively higher doses of a chemical till at least half of a group dies. The OECD had to wait until 2000 for the EPA to acquiesce so that it could abandon the test, despite other countries rejecting it years ago.

The EPA's reticence stems from plain old conservatism, lack of organisation and mistrust of anything new, according to Richard Hill, who represents the agency at OECD meetings. Many animal tests have become entrenched through years of use, he says, and there is no quick process for bringing in alternatives.

Hill says that things are finally changing, however, mainly thanks to the government-wide Interagency Coordinating Committee on the Validation of Alternative Methods (ICCVAM), created in 1997 to bring federal agencies together. "For the first time, we had a forum where we could talk about what's good and what's bad about old and new methods," says Hill. For example, the EPA organised a workshop in February [2002] to teach industry and government researchers how to switch from the LD50 test to alternatives that use only up to nine animals instead of 25.

But critics claim this is not enough, and say that US law must change to match the legislation in Europe that forces labs to use alternatives to animal tests whenever they become available. Worse, says McArdle, changes to the US Animal Welfare Act [in May 2002] have relegated rats, mice and birds to "nonanimal" status, meaning they're no longer protected by welfare legislation, despite being used in 95 per cent of animal experiments.

Hill admits there is less impetus in the US than in Europe to push for in vitro testing. But he says that there are simply no alternatives in some cases. For example, in vitro tests to see whether chemicals harm the eye have consistently failed to measure up. "It doesn't matter if you want in vitro methods, because sometimes you can't have them—the science is not there," he says.

And there's the rub. "The bottom line is that new methods must provide at least the equivalent protection for human health and the environment, or preferably better," says Fred Stokes of ICCVAM. "It would be irresponsible to adopt new methods that don't do this." Balls agrees. "If you replace tests, they'd better work or you'd have another thalidomide on your hands."

More to Be Done

But overall there is wide agreement that much more can be done to lessen the pain and suffering caused to animals in medical research. Animal welfare proponents say one reason why the US is dragging its feet is that the American public is less vociferous than people in Europe when it comes to opposing animal testing. "It's much more in-your-face over here," says Bob Combes, scientific director of Britain's Fund for the Replacement of Animals in Medical Experiments (FRAME).

The other major barrier to alternatives is a lack of funding, and that's not confined to the US. There is a classic mismatch between what the public demands and what governments provide, says Herman Koeter, a stalwart who for years has run the OECD's programme to provide guidelines on animal testing.

He says the number of proposed nonanimal tests is drying up, primarily because regulatory authorities and government labs aren't developing new ones. Most new alternative tests are being developed by industry or private charities.

Langley says that the British government allocates less than 60,000 [pounds] a year for researching alternatives. "That's a joke," she says. It is also a false economy, because alternative tests are faster and cheaper than those using animals.

Taking Animal Welfare Seriously

Yet it looks as though governments are at last taking animal welfare seriously, and a significant number of the old tests may finally be on the way out. [In 2002], animal welfare charities have been allowed to sit in on the OECD meetings for the first time, through an umbrella group called the International Council on Animal Protection in the OECD. "It's fantastic," says Langley. "We don't have a vote, but I think it'll help change the tone and background against which these decisions are taken," she says.

[In 2002], the council secured a major victory, convincing member states to prioritise the validation of a new technique for assessing how carcinogenic chemicals are. Traditionally, between 500 and 1000 rodents are needed to test each chemical, over a period of two years. The new test uses cells from Syrian hamster embryos and gives a result in days. "It would save a massive amount of money and thousands of animals," says Langley.

And [in July 2002], in response to the widespread desire for change, animal welfare groups, politicians, and scientists gathered at a European Commission meeting to galvanise support for other alternative tests.

"The groundswell towards non-animal testing has been building for decades and is beginning to accelerate," says Langley. All that's now required is money to support the new ideology.

ALTERNATIVES TO ANIMAL DISSECTION IN THE CLASSROOM

David J. Hoff

High school students who do not want to dissect animals in biology class now have an option, writes David J. Hoff. Advances in technology now allow students with ethical objections to dissection to use tools such as CD-ROM to participate in virtual dissection, Hoff explains. However, not all teachers are sold on alternatives to dissection, Hoff reveals. They believe that only a real animal can provide students with the full educational benefits of studying internal anatomy. Nevertheless, some states now have laws that require informing students that dissection will be required before they enroll in a class, the author relates, and some laws additionally require offering students an alternative. While Hoff reports that many students are willing to dissect, technology allows students to learn without compromising their ethics. Hoff is an assistant editor for *Education Week*.

When Lauren S. Skaskiw refused to dissect a fetal pig [in 2000], she learned a lesson in biology—and civics.

The junior at Woodstock Union High School in eastern Vermont managed to earn her class's highest grade on the dissection-lab report, even though she participated "virtually" through a CD-ROM and a model of a pig.

She then embarked on a campaign that took her to the Statehouse in Montpelier, back to a school board meeting, and to a local committee meeting that she hopes will result in a new policy requiring the school to offer alternatives to dissecting a specimen.

"High school students are mature enough to know whether they want to dissect an animal," Ms. Skaskiw said in [an] interview. "My classmates doing the dissection were not learning things. They were playing with the organs. They were getting sick because of the smell or because they were grossed out about what they were doing."

She is part of a new generation of animal-rights activists who demand that they be exempted from what has long been a rite of passage for high school biology students: the cutting open and exploration of a once-living animal.

Aided by new technologies that mimic the dissection experience and laws in some states that require schools to find substitutes for such class assignments, students like Ms. Skaskiw often are successful in learning the material and achieving high grades without violating their principles or taking part in an experiment they find offensive.

But biology teachers question whether those who pursue an alternative path get a full understanding of how an animal's organs, tissues, and nervous systems cooperate to keep it alive.

"There can be good learning experiences from computers and other technologies," said Wayne Carley, the executive director of the National Association of Biology Teachers, a Reston, Va.-based membership group. "But they do not replace and are not equivalent to dissecting a real animal. If you want to teach a kid how a muscle works, you have to let that kid get inside the animal and tug and pull."

Few Objectors

In the past decade, animal-rights proponents—better known for their protests against fur coats and the use of living animals in research labs—have also focused their attention on getting rid of dissection in schools.

The groups, for example, contend that the practice of dissecting frogs—the most common species used in introductory biology—is one reason for declining populations of the amphibians in some regions. While acid rain and pesticides probably have more to do with the diminishing frog numbers, capturing frogs for use in high school biology classes can't help the species, animal-rights activists say.

Frogs, as well as cats, rats, and turtles, are stored in inhumane conditions by companies that supply schools with the specimens, according to the Humane Society of the United States, a leading voice against dissections.

While the Humane Society and the activist group People for the Ethical Treatment of Animals have encouraged opposition to dissection, Mr. Carley and biology teachers say they see little protest and few students opting out.

"There's always a couple, every now and then," said Philip J. McCrea, a biology teacher at New Trier High School in Winnetka, Ill., and a former biology-association president. "As soon as one kid makes a big, verbal statement, then suddenly three or four others do."

Mr. McCrea said only one student of his refused to dissect a fetal pig in an introductory-biology class [in 2001]. While her classmates worked on teams poking and probing the specimens, she explored the animal in a CD-ROM produced by ScienceWorks, a Winston-Salem, N.C., company that produces simulated dissections of pigs, frogs, earthworms, perch, and crayfish.

For the exam on the unit, Mr. McCrea showed the girl pictures of a pig's organs and asked her to identify them. Her classmates were assessed on their ability to identify the organs of an actual specimen. While the objector aced the exam, her teacher said he isn't convinced she understood the biology of the pig as well as others in the class.

"She didn't get the intricacy of seeing the tissues and how they fit together, and how all the membranes hold it together," Mr. McCrea said. "She can't get that by looking at a picture."

While the student may not have acquired as complete a scientific lesson, she did learn one about the value of life, according to the Humane Society. Live dissections, the group argues, treat animals as "expendable" and teach disrespect for their lives.

Adequate Alternatives

Animal-rights advocates say dissecting by video or computer screen can produce a high-quality learning experience.

"Most of the time, [alternative dissections] work out very well," said Cheryl L. Ross, a research assistant for animal-research issues at the Humane Society, a Washington-based animal-welfare group. "We find that a lot of students learned exactly the same thing—or even more."

In the Digital Frog 2, a CD-ROM produced by a Puslinch, Ontario, company, students select what portion of the amphibian to dissect. They are prompted to draw a line connecting two dots on the screen to mimic a scalpel cut. When they're done, they watch a video of an actual dissection, with a narrator describing what is happening.

The CD-ROM also includes sections that ask the user to highlight organs or body parts on the screen.

The Humane Society runs a loan program from which teachers and students can borrow materials for students who don't want to dissect a specimen. Products such as Digital Frog 2 "give a good overall view of dissection" without forcing students to do something they might find objectionable, Ms. Ross said.

But many teachers believe the computer models work best as supplements.

Michael D. Nassise, the science-department chairman at Stoughton High School in Massachusetts, about 20 miles south of Boston, said a student can refer to the computer model if he or she makes an incorrect cut and ruins an organ.

"It supplements the actual dissection very, very nicely," said Mr. Nassise, who notes that he makes accommodations for students who don't want to dissect.

Allowing Students to Have a Choice

Use of such simulations as alternatives, not just supplements, may rise as the Humane Society and other organizations push for laws requiring that students be informed they will have to cut up an animal specimen before they sign up for a course, and compelling schools to substitute other activities.

"There's no reason to [require a dissection], because there are alternatives that are as good and often superior to dissecting a specimen," argued Theodora Capaldo, the president of the Ethical Science and Education Coalition, a Boston group lobbying for a Massachusetts law requiring schools to give students such choices.

Such laws started to go on the books in the late 1980s, following a suburban Los Angeles girl's lawsuit to get her grade raised after she produced a report based on a computer-based dissection. Because Jenifer Graham refused to cut open an animal, her teacher had lowered her grade from an A to a B.

California and seven other states require schools to tell students dissection will be part of a biology course. California, Illinois, Maryland, Pennsylvania, and Rhode Island also mandate that schools offer an alternative to conducting an actual dissection, according to the Ethical Science and Education Coalition.

The Illinois and Maryland laws have had little impact, according to biology teachers in those states, because most schools were already offering alternatives.

"The legislation has not caused a rampant rush to avoid dissection," said Dale E. Peters, a biology teacher at Urbana High School in Maryland and the president of the Maryland Association of Science Teachers. "For the most part, it's been a nonissue."

In Illinois, most urban and suburban districts have long provided backup measures, Mr. McCrea said. Some rural districts may have to scramble to comply with the . . . law, but in those areas, students who have grown up on or near farms are usually comfortable around dead animals, he remarked.

But such laws are necessary because teachers are authority figures whom most students don't want to challenge, according to the Ethical Science and Education Coalition. Even if schools give students a choice, "you will get the occasional teacher who will duke it out with a student," Ms. Capaldo said.

Many teachers, though, say they willingly allow students to opt out of dissecting an animal, especially in an introductory course.

It's a different story for students in Advanced Placement biology or an anatomy course. They need the experience of an actual dissection because many will be pursuing science as a major in college or as a career, said James E. Slouf, the science chairman at Downers Grove South High School in suburban Chicago.

Students who sign up for those courses know ahead of time that they will be expected to dissect animals, Mr. Slouf said.

After being told that her class would be investigating a fetal pig, Ms. Skaskiw, the Vermont student, approached her teacher and announced her objections. The teacher urged her to change her mind, telling her that she would not have the same in-depth learning experience with a computer program.

Ms. Skaskiw appealed to the science-department chairman, who granted her an exemption from the live dissection, she said.

She and her mother researched other states' laws and started lobbying the Vermont legislature. She worked closely with state Sen. Cheryl Rivers, the Democrat who chairs the Senate education committee, to draft a bill that was introduced in [the 2001] session.

Ms. Skaskiw circulated petitions supporting the bill, solicited letters endorsing it, and produced a video with testimonials for her cause. [In the] spring, she traveled to Montpelier to testify before Ms. Rivers' committee.

In the end, lawmakers told the 17-year-old that the bill wouldn't pass the legislature [in 2001] because they saw it as an intrusion on local decisionmaking, Ms. Skaskiw said. At her urging, however, the school board of the 1,400-student Windsor Central district told the high school's principal and science teachers to craft a policy that encompasses the intent of the bill.

When the school committee meets, Ms. Skaskiw plans to be there to ensure that it fully protects the rights of students who don't want a hands-on science lesson in cutting open an animal.

ORGANIZATIONS TO CONTACT

The editors have compiled the following list of organizations concerned with the issues presented in this book. The descriptions are derived from materials provided by the organizations. All have publications or information available for interested readers. The list was compiled on the date of publication of the present volume; the information provided here may change. Be aware that many organizations take several weeks or longer to respond to inquiries, so allow as much time as possible.

American Anti-Vivisection Society (AAVS)
801 Old York Rd., Suite 204, Jenkintown, PA 19046
(215) 887-0816 • fax: (215) 887-2088
Web site: www.aavs.org

AAVS advocates the abolition of vivisection, opposes all types of experiments on living animals, and sponsors research on alternatives to these methods. The society produces videos and publishes numerous brochures, including *Vivisection and Dissection and the Classroom: A Guide to Conscientious Objection* and the bimonthly *AV Magazine*.

American Association for Laboratory Animal Science (AALAS)
9190 Crestwyn Hills Dr., Memphis, TN 38125
(901) 754-8620 • fax: (901) 759-5849
Web site: www.aalas.org

AALAS is a professional nonprofit association concerned with the production, care, and study of animals used in biomedical research. The association provides a medium for the exchange of scientific information on all phases of laboratory animal care and use through educational activities, publications, and certification programs.

Americans for Medical Progress (AMP)
421 King St., Suite 401, Alexandria, VA 22314
(703) 836-9595 • fax: (703) 836-9594
e-mail: info@amprogress.org • Web site: www.amprogress.org

AMP is a nonprofit organization working to raise public awareness concerning the use of animals in research in order to ensure that scientists and doctors have the freedom and resources necessary to pursue their research. Through newspaper and magazine articles, broadcast debates, and public education materials, AMP exposes what it characterizes as the misinformation of the animal rights movement.

American Society for the Prevention of Cruelty to Animals (ASPCA)
424 E. Ninety-second St., New York, NY 10128-6804
(212) 876-7700
e-mail: information@aspca.org • Web site: www.aspca.org

Established in 1866, the nonprofit ASPCA is the oldest animal humane organization in the United States. Its mission is to provide effective means for preventing animal cruelty. ASPCA offers national programs in humane education, public awareness, government advocacy, shelter support, and animal medical services and placement. The society's publications include numerous press releases, online brochures, and the magazine *ASPCA Animal Watch*.

Animal Aid
The Old Chapel, Bradford St., Tonbridge, Kent, TN9 1AW United Kingdom
44 (0) 1732 364546 • fax: 44 (0) 1732 366533
e-mail: info@animalaid.org.uk • Web site: www.animalaid.org.uk
Animal Aid investigates and exposes animal cruelty. The organization stages street protests and education tours and publishes educational packs for schools and colleges.

Animal Alliance of Canada
221 Broadway Ave., Suite 101, Toronto, ON M4M 2G3 Canada
(416) 462-9541 • fax: (416) 462-9647
Web site: www.animalalliance.ca
The Animal Alliance of Canada is an animal rights advocacy and education group that focuses on local, regional, national, and international issues concerning the treatment of animals by humans. The alliance acts through research, investigation, education, advocacy, and legislation. Publications include fact sheets, legislative updates, editorials, and the newsletter *Take Action*.

Animal Welfare Institute (AWI)
PO Box 3650, Washington, DC 20007
(202) 337-2332 • fax: (202) 338-9478
e-mail: awi@animalwelfare.org • Web site: www.animalwelfare.org
Founded in 1951, the AWI is a nonprofit charitable organization working to reduce pain and fear inflicted on animals by humans. AWI believes in the humane treatment of laboratory animals and the development and use of nonanimal testing methods. It encourages humane science teaching and prevention of painful experiments on animals by high school students. In addition to publishing *AWI Quarterly*, the institute also offers numerous books, pamphlets, and online articles.

Fund for Animals
200 W. Fifty-seventh St., New York, NY 10019
(212) 246-2096 • fax: (212) 246-2633
e-mail: fundinfo@fund.org • Web site: www.fund.org
The Fund for Animals was founded in 1967 by prominent author and animal advocate Cleveland Amory. It is one of the largest and most active organizations working for the welfare of both wild and domestic animals throughout the world. The fund works on education, legislation, litigation, and hands-on animal care.

Incurably Ill for Animal Research
PO Box 27454, Lansing, MI 48909
(517) 887-1141
e-mail: info@iifar.org • Web site: www.iifar.org
This organization consists of people who have incurable diseases and are concerned that the use of animals in medical research will be stopped or severely limited by animal rights activists, thus delaying or preventing the development of new cures. It publishes the monthly *Bulletin* and a quarterly newsletter.

In Defense of Animals
131 Camino Alto, Suite E, Mill Valley, CA 94941
(415) 388-9641 • fax: (415) 388-0388
e-mail: ida@idausa.org • Web site: www.idausa.org

In Defense of Animals is a nonprofit organization established in 1983 that works to end the institutional exploitation and abuse of laboratory animals. The organization publishes fact sheets and brochures on animal abuse in the laboratory and on how to live a cruelty-free lifestyle.

Institute for In Vitro Sciences (IIVS)
21 Firstfield Rd., Suite 220, Gaithersburg, MD 20878
(301) 947-6523 • fax: (301) 947-6538
Web site: www.iivs.org

The institute is a nonprofit technology-driven foundation for the advancement of alternative methods to animal testing. Its mission is to facilitate the replacement of animal testing through the use of in vitro technology, conduct in vitro testing for industry and government, and provide educational and technical resources to public and private sectors.

National Animal Interest Alliance (NAIA)
PO Box 66579, Portland, OR 97290
(503) 761-1139
e-mail: NAIA@naiaonline.org • Web site: www.naiaonline.org

NAIA is an association of business, agricultural, scientific, and recreational interests formed to protect and promote humane practices and relationships between people and animals. NAIA provides the network necessary for diverse animal rights groups to communicate with one another, describe the nature and value of their work, clarify animal rights misinformation, and educate each other and the public about what they do. NAIA serves as a clearinghouse for information and as an access point for subject matter experts, keynote speakers, and issue analysts. The alliance also publishes the bimonthly newspaper *NAIA News*.

National Cattlemen's Beef Association (NCBA)
1301 Pennsylvania Ave. NW, Suite 300, Washington, DC 20004
(202) 347-0228 • fax: (202) 638-0607
Web site: www.beef.org

NCBA was founded in 1898 as a marketing and trade organization for America's cattle farmers and ranchers. The association oversees beef production to ensure the safety and quality of America's beef supply. NCBA works to encourage the humane treatment of farm animals, the wise stewardship of natural resources, and the implementation of good husbandry practices. The association's publications include brochures on beef preparation and nutrition, as well as *National Cattlemen Magazine*.

People for the Ethical Treatment of Animals (PETA)
501 Front St., Norfolk, VA 23510
(757) 622-7382 • fax: (757) 622-0457
e-mail: peta@norfolk.infini.net • Web site: www.peta.org

An international animal rights organization, PETA is dedicated to establishing and protecting the rights of all animals. It focuses on four areas: factory farms, research laboratories, the fur trade, and the entertainment industry. PETA promotes public education, cruelty investigations, animal rescue, celebrity involvement, and legislative action. It produces numerous videos and publishes *Animal Times, Grrr!* and various fact sheets, brochures, and flyers.

Psychologists for the Ethical Treatment of Animals (PSYETA)
403 McCauley St., PO Box 1297, Washington Grove, MD 20880
(301) 565-4167
Web site: www.psyeta.org

PSYETA seeks to ensure proper treatment of animals used in psychological research and education and urges revision of curricula to include ethical issues in the treatment of animals. It developed a tool to measure the invasiveness or severity of animal experiments. Its publications include the book *Animal Models of Human Psychology* and the journals *Society and Animals* and *Journal of Applied Animal Welfare Science.*

Vegan Action
PO Box 4288, Richmond, VA 23220
(804) 502-8736 • fax: (804) 355-7934
e-mail: information@vegan.org • Web site: www.vegan.org

Vegan Action is a nonproft organization dedicated to helping animals, the environment, and human health by educating the public about the benefits of a vegan lifestyle and encouraging the spread of vegan food options through public outreach campaigns. It publishes the monthly electronic newsletter *Vegan Action News.*

BIBLIOGRAPHY

Books

Paola Cavalieri	*The Animal Question: Why Nonhuman Animals Deserve Human Rights.* Oxford, UK: Oxford University Press, 2001.
Tony Gilland	*Animal Experimentation: Good or Bad?* London. Hodder & Stoughton, 2002.
Jane Goodall and Mark Bekoff	*The Ten Trusts: What We Must Do to Care for the Animals We Love.* New York: HarperCollins, 2002.
C. Ray Greek and Jean Swingle Greek	*Sacred Cows and Golden Geese: The Human Cost of Experiments on Animals.* New York: Continuum, 2000.
Anita Guerrini	*Experimenting with Humans and Animals.* Baltimore: Johns Hopkins University Press, 2003.
John M. Kistler	*People Promoting and People Opposing Animal Rights: In Their Own Words.* Westport, CT: Greenwood Press, 2002.
Tibor R. Machan	*Putting Humans First: Why We Are Nature's Favorite.* Lanham, MD: Rowman & Littlefield, 2004.
Lyle Munro	*Compassionate Beasts: The Quest for Animal Rights.* Westport, CT: Praeger, 2001.
Ellen Frankel Paul et al.	*Why Animal Experimentation Matters: The Use of Animals in Medical Research.* Washington, DC: Social Philosophy Policy Center, 2001.
David Perkins	*Romanticism and Animal Rights.* Cambridge, UK: Cambridge University Press, 2003.
Tom Regan	*Empty Cages: Facing the Challenge of Animal Rights.* Lanham, MD: Rowman & Littlefield, 2004.
Steve F. Sapontzis, ed.	*Food for Thought: The Debate over Eating Meat.* Amherst, NY: Prometheus, 2004.
Matthew Scully	*Dominion: The Power of Man, the Suffering of Animals, and the Call to Mercy.* New York: St. Martin's Press, 2002.
Peter Singer	*Animal Liberation.* New York: New York Review of Books, 1975.
Angus Taylor	*Animals and Ethics: An Overview of the Philosophical Debate.* Peterborough, Ontario: Broadview, 2003.
Gary E. Varner	*In Nature's Interests? Interests, Animal Rights, and Environmental Ethics.* Oxford, UK: Oxford University Press, 2002.

Periodicals

David R. Carlin   "Rights, Animal and Human," *First Things*, August/ September 2000.

Clifton Coles   "Legal Personhood for Animals," *Futurist*, September 2002.

Charles Colson and Anne Morse   "Taming Beasts," *Christianity Today*, April 2003.

Economist   "It's a Dog's Life," December 21, 2002.

Barney Gimbel   "Filing for Fido," *Newsweek*, September 1, 2003.

Jonah Goldberg   "Soy Vey!" *National Review*, February 10, 2003.

Frederick K. Goodwin and Adrian R. Morrison   "Science and Self-Doubt," *Reason*, October 2000.

Catherine Gourley   "Animal Welfare or Animal Rights?" *Writing*, January 2002.

Emily Green   "The High Price of Cheap Food," *Los Angeles Times*, January 21, 2004.

Christopher Hitchens   "Political Animals," *Atlantic Monthly*, November 2002.

Steven Menashi   "Humans, Animals, and the Human Animal," *Policy Review*, February/March 2003.

Ed Metcalfe   "The Pig Issue," *Ecologist*, December 2001–January 2002.

Marlys Miller   "Oh, What a Tangled Web They Weave," *AgriMarketing*, July/August 2003.

Gerald Mizejewski   "Animal Crackers," *Insight on the News*, April 16, 2001.

Eric Moore   "The Case for Unequal Animal Rights," *Environmental Ethics*, September 2002.

Adrian R. Morrison   "Making Choices in the Laboratory," *Society*, September/October 2002.

Doug Moss   "He Ain't Hairy, He's My Brother," *E Magazine*, March/ April 2003.

Jim Motavalli   "Across the Great Divide," *E Magazine*, January/ February 2002.

Richard John Neuhaus   "The Quality of Mercy," *National Review*, December 31, 2002.

David S. Oderberg   "The Illusion of Animal Rights," *Human Life Review*, Spring/Summer 2000.

Michele Orecklin   "When Animal Lovers Attack," *Time*, November 25, 2002.

Michael Pollan   "An Animal's Place," *New York Times Magazine*, November 10, 2002.

Janet Raloff   "Of Rats, Mice, and Birds," *Science News*, November 18, 2000.

Megan Rooney   "Getting Lambs off the Hook," *Chronicle of Higher Education*, June 20, 2003.

Julian Sanchez   "Meat Markets," *Reason*, October 2003.

Niall Shanks   "Animal Rights in the Light of Animal Cognition," *Social Alternatives*, Summer 2003.

Ron Southwick   "Fighting for Research on Animals," *Chronicle of Higher Education*, April 12, 2002.

Steven L. Teitelbaum   "Animal Rights Pressure on Scientists," *Science*, November 22, 2002.

INDEX